Data Warehousing and Data Mining Techniques for Cyber Security

Advances in Information Security

Sushil Jajodia
Consulting Editor
Center for Secure Information Systems
George Mason University
Fairfax, VA 22030-4444
email: jajodia@gmu.edu

The goals of the Springer International Series on ADVANCES IN INFORMATION SECURITY are, one, to establish the state of the art of, and set the course for future research in information security and, two, to serve as a central reference source for advanced and timely topics in information security research and development. The scope of this series includes all aspects of computer and network security and related areas such as fault tolerance and software assurance.

ADVANCES IN INFORMATION SECURITY aims to publish thorough and cohesive overviews of specific topics in information security, as well as works that are larger in scope or that contain more detailed background information than can be accommodated in shorter survey articles. The series also serves as a forum for topics that may not have reached a level of maturity to warrant a comprehensive textbook treatment.

Researchers, as well as developers, are encouraged to contact Professor Sushil Jajodia with ideas for books under this series.

Additional titles in the series:

Additional information about this series can be obtained from
http://www.springer.com

Data Warehousing and Data Mining Techniques for Cyber Security

by

Anoop Singhal
NIST, Computer Security Division
USA

 Springer

Anoop Singhal
NIST, Computer Security Division
National Institute of Standards and Tech
Gaithersburg MD 20899
psinghal@nist.gov

Library of Congress Control Number: 2006934579

Data Warehousing and Data Mining Techniques for Cyber Security
 by Anoop Singhal

ISBN-10: 0-387-26409-4
ISBN-13: 978-0-387-26409-7
e-ISBN-10: 0-387-47653-9
e-ISBN-13: 978-0-387-47653-7

Printed on acid-free paper.

Printed in the United States of America.

9 8 7 6 5 4 3 2 1

springer.com

PREFACE

The fast growing, tremendous amount of data, collected and stored in large databases has far exceeded our human ability to comprehend it without proper tools. There is a critical need of data analysis systems that can automatically analyze the data, summarize it and predict future trends. Data warehousing and data mining provide techniques for collecting information from distributed databases and then performing data analysis.

In the modern age of Internet connectivity, concerns about denial of service attacks, computer viruses and worms have become very important. There are a number of challenges in dealing with cyber security. First, the amount of data generated from monitoring devices is so large that it is humanly impossible to analyze it. Second, the importance of cyber security to safeguard the country's Critical Infrastructures requires new techniques to detect attacks and discover the vulnerabilities. The focus of this book is to provide information about how data warehousing and data mining techniques can be used to improve cyber security.

OBJECTIVES

The objective of this book is to contribute to the discipline of Security Informatics. It provides a discussion on topics that intersect the area of Cyber Security and Data Mining. Many of you want to study this topic: College and University students, computer professionals, IT managers and users of computer systems. The book will provide the depth and breadth that most readers want to learn about techniques to improve cyber security.

INTENDED AUDIENCE

What background should you have to appreciate this book? Someone who has an advanced undergraduate or graduate degree in computer science certainly has that background. We also provide enough background material in the preliminary chapters so that the reader can follow the concepts described in the later chapters.

PLAN OF THE BOOK

Chapter 1: Introduction to Data Warehousing and Data Mining

This chapter introduces the concepts and basic vocabulary of data warehousing and data mining.

Chapter 2: Introduction to Cyber Security

This chapter discusses the basic concepts of security in networks, denial of service attacks, network security controls, computer virus and worms

Chapter 3: Intrusion Detection Systems

This chapter provides an overview of the state of art in Intrusion Detection Systems and their shortcomings.

Chapter 4: Data Mining for Intrusion Detection

It shows how data mining techniques can be applied to Intrusion Detection. It gives a survey of different research projects in this area and possible directions for future research.

Chapter 5: Data Modeling and Data Warehousing to Improve IDS

This chapter demonstrates how a multidimensional data model can be used to do network security analysis and detect denial of service attacks. These techniques have been implemented in a prototype system that is being successfully used at Army Research Labs. This system has helped the security analyst in detecting intrusions and in historical data analysis for generating reports on trend analysis.

Chapter 6: MINDS: Architecture and Design

It provides an overview of the Minnesota Intrusion Detection System (MINDS) that uses a set of data mining techniques to address different aspects of cyber security.

Chapter 7: Discovering Novel Strategies from INFOSEC Alerts

This chapter discusses an advanced correlation system that can reduce alarm redundancy and provide information on attack scenarios and high level attack strategies for large networks.

ACKNOWLEDGEMENTS

This book is the result of hard work by many people. First, I would like to thank Prof. Vipin Kumar and Prof. Wenke Lee for contributing two chapters in this book. I would also like to thank Melissa, Susan and Sharon of Springer for their continuous support through out this project. It is also my pleasure to thank George Mason University, Army Research Labs and National Institute of Standards and Technology (NIST) for supporting my research on cyber security.

Authors are products of their environment. I had good education and I think it is important to pass it along to others. I would like to thank my parents for providing me good education and the inspiration to write this book.

-Anoop Singhal

TABLE OF CONTENTS

Chapter 1

AN OVERVIEW OF DATA WAREHOUSE, OLAP AND DATA MINING TECHNOLOGY

Anoop Singhal

Abstract: In this chapter, a summary of Data Warehousing, OLAP and Data Mining Technology is provided. The technology to build Data Analysis Application for Network/Web services is also described

Key words: STAR Schema, Indexing, Association Analysis, Clustering

1. MOTIVATION FOR A DATA WAREHOUSE

Data warehousing (DW) encompasses algorithms and tools for bringing together data from distributed information repositories into a single repository that can be suitable for data analysis [13]. Recent progress in scientific and engineering applications has accumulated huge volumes of data. The fast growing, tremendous amount of data, collected and stored in large databases has far exceeded our human ability to comprehend it without proper tools. It is estimated that the total database size for a retail store chain such as Walmart will exceed 1 Petabyte (1K Terabyte) by 2005. Similarly, the scope, coverage and volume of digital geographic data sets and multidimensional data has grown rapidly in recent years. These data sets include digital data of all sorts created and disseminated by government and private agencies on land use, climate data and vast amounts of data acquired through remote sensing systems and other monitoring devices [16], [18]. It is estimated that multimedia data is growing at about 70% per year. Therefore, there is a critical need of data analysis systems that can automatically analyze the data, to summarize it and predict future trends. Data warehousing is a necessary technology for collecting information from distributed databases and then performing data analysis [1], [2], [3], and [4].

Data warehousing is an enabling technology for data analysis applications in the area of retail, finance, telecommunication/Web services and bio-informatics. For example, a retail store chain such as Walmart is interested in integrating data from its inventory database, sales database from different stores in different locations, and its promotions from various departments. The store chain executives could then 1) determine how sales trend differ across regions of the country 2) correlate its inventory with current sales and ensure that each store's inventory is replaced to keep up with the sales 3) analyze which promotions are leading to increases product sales. Data warehousing can also be used in telecommunication/Web services applications for collecting the usage information and then identify usage patterns, catch fraudulent activities, make better use of resources and improve the quality of service. In the area of bio-informatics, the integration of distributed genome databases becomes an important task for systematic and coordinated analysis of DNA databases. Data warehousing techniques will help in integration of genetic data and construction of data warehouses for genetic data analysis. Therefore, analytical processing that involves complex data analysis (usually termed as decision support) is one of the primary uses of data warehouses [14].

The commercial benefit of Data Warehousing is to provide tools for business executives to systematically organize, understand and use the data for strategic decisions. In this paper, we motivate the concept of a data warehouse, provide a general architecture of data warehouse and data mining systems, discuss some of the research issues and provide information on commercial systems and tools that are available in the market.

Some of the key features of a data warehouse (DW) are as follows.

1. Subject Oriented: The data in a data warehouse is organized around major subjects such as customer, supplier and sales. It focuses on modeling data for decision making.
2. Integration: It is constructed by integrating multiple heterogeneous sources such as RDBMS, flat files and OLTP records.
3. Time Variant: Data is stored to provide information from a historical perspective.

The data warehouse is physically separate from the OLTP databases due to the following reasons:

1. Application databases are 3NF optimized for transaction response time and throughput. OLAP databases are market oriented and optimized for data analysis by managers and executives.
2. OLTP systems focus on current data without referring to historical data. OLAP deals with historical data, originating from multiple organizations.
3. The access pattern for OLTP applications consists of short, atomic transactions where as OLAP applications are primarily read only transactions that perform complex queries.

These characteristics differentiate data warehouse applications from OLTP applications and they require different DBMS design and implementation techniques. Clearly, running data analysis queries over globally distributed databases is likely to be excruciatingly slow. The natural solution is to create a centralized repository of all data i.e. a data warehouse. Therefore, the desire to do data analysis and data mining is a strong motivation for building a data warehouse.

This chapter is organized as follows. Section 2 discusses the multi-dimensional data model and section 3 discusses the data warehouse architecture. Section 4 discusses the implementation techniques and section 5 presents commercial tools available to implement data warehouse systems. Section 6 discusses the concepts of Data Mining and applications of data mining. Section 7 presents a Data Analysis Application using Data Warehousing technology that the authors designed and implemented for AT&T Business Services. This section also discusses some open research problems in this area. Finally section 8 provides the conclusions.

2.   A MULTIDIMENSIONAL DATA MODEL

Data Warehouse uses a data model that is based on a *multidimensional data model*. This model is also known as a *data cube* which allows data to be modeled and viewed in multiple dimensions. *Dimensions* are the different perspectives for an entity that an organization is interested in. For example, a store will create a *sales* data warehouse in order to keep track of the store' sales with respect to different dimensions such as *time, branch, and location.* "Sales" is an example of a central theme around which the data model is organized. This central theme is also referred as a *fact table.* *Facts* are numerical measures and they can be thought of as quantities by which we want to analyze relationships between dimensions. Examples of facts are *dollars_sold, units_sold* and so on. The *fact table* contains the names of the facts as well as keys to each of the related dimension tables.

The entity-relationship data model is commonly used in the design of relational databases. However, such a schema is not appropriate for a data warehouse. A data warehouse requires a concise, subject oriented schema that facilitates on-line data analysis. The most popular data model for a data warehouse is a *multidimensional model.* Such a model can exist in the form of a *star schema.* The star schema consists of the following.
1. A large central table (fact table) containing the bulk of data.
2. A set of smaller *dimension tables* one for each dimension.

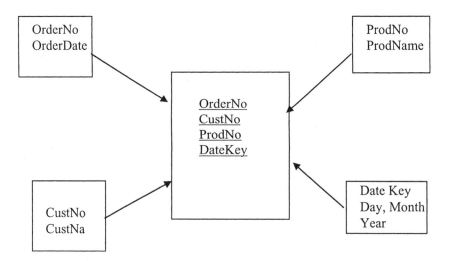

Figure 1: A Star Schema

The schema resembles a star, with the dimension tables displayed in a radial pattern around the central fact table. An example of a sales table and the corresponding star schema is shown in the figure 1. For each dimension, the set of associated values can be structured as a hierarchy. For example, cities belong to states and states belong to countries. Similarly, dates belong to weeks that belong to months and quarters/years. The hierarchies are shown in figure 2.

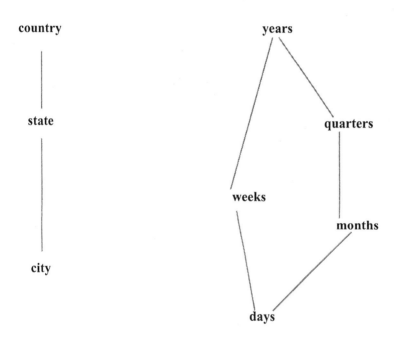

Figure 2: Concept Hierarchy

In data warehousing, there is a distinction between a data warehouse and a data mart. A data warehouse collects information about subjects that span the entire organization such as customers, items, sales and personnel. Therefore, the scope of a data warehouse is *enterprise wide*. A data mart on the other hand is a subset of the data warehouse that focuses on selected subjects and is therefore limited in size. For example, there can be a data mart for sales information another data mart for inventory information.

3.   DATA WAREHOUSE ARCHITECTURE

Figure 3 shows the architecture of a Data Warehouse system. Data warehouses often use three tier architecture.

1. The first level is a warehouse database server that is a relational database system. Data from operational databases and other external sources is extracted, transformed and loaded into the database server.
2. Middle tier is an OLAP server that is implemented using one of the following two methods. The first method is to use a relational OLAP model that is an extension of RDBMS technology. The second method is to use a multidimensional OLAP model that uses a special purpose server to implement the multidimensional data model and operations.
3. Top tier is a client which contains querying, reporting and analysis tools.

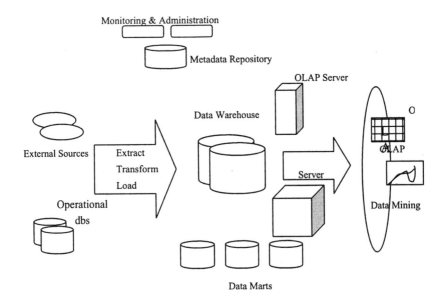

Figure 3: Architecture of a Data Warehouse System

4.   DATA WAREHOUSE IMPLEMENTATION

Data warehouses contain huge volumes of data. Users demand that decision support queries be answered in the order of seconds. Therefore, it is critical for data warehouse systems to support highly efficient cube computation techniques and query processing techniques. At the core of multidimensional analysis is the efficient computation of aggregations across many sets of dimensions. These aggregations are referred to as **group-by**. Some examples of "group-by" are

1. Compute the sum of sales, grouping by item and city.
2. Compute the sum of sales, grouping by item.

Another use of aggregation is to summarize at different levels of a dimension hierarchy. If we are given total sales per city, we can aggregate on the location dimension to obtain sales per state. This operation is called **roll-up** in the OLAP literature. The inverse of roll-up is **drill-down:** given total sales by state, we can ask for a more detailed presentation by drilling down on location. Another common operation is **pivoting**. Consider a tabular presentation of Sales information. If we pivot it on the Location and Time dimensions, we obtain a table of total sales for each location for each time value. The time dimension is very important for OLAP. Typical queries are

- Find total sales by month
- Find total sales by month for each city
- Find the percentage change in total monthly sales

The OLAP framework makes it convenient to implement a broad class of queries. It also gives the following catchy names:

- **Slicing:** a data set amounts to an equality selection on one or more dimensions
- **Dicing:** a data set amounts to a range selection.

4.1   Indexing of OLAP Data

To facilitate efficient data accessing, most data warehouse systems support index structures and materialized views. Two indexing techniques that are popular for OLAP data are *bitmap indexing* and *join indexing*.

4.1.1   Bitmap indexing

The bitmap indexing allows for quick searching in data cubes. In the bit map index for a given attribute, there is a distinct bit vector, Bv, for each value v in the domain of the attribute. If the domain for the attribute consists of n values, then n bits are needed for each entry in the bitmap index.

OLAP =
On Line Analytical
Processing

4.1.2   Join indexing

Consider 2 relations R(RID, A) and S(B, RID) that join on attributes A and B. Then the join index record contains the pair (RID, SID) where RID and SID are record identifiers from the R and S relations. The advantage of join index records is that they can identify joinable tuples without performing costly join operations. Join indexing is especially useful in the star schema model to join the fact table with the corresponding dimension table.

4.2   Metadata Repository

Metadata is data about data. A meta data repository contains the following information.

1. A description of the structure of data warehouse that includes the schema, views and dimensions.
2. Operations metadata that includes data lineage (history of data and the sequence of transformations applied to it).
3. The algorithms used for summarization.
4. The mappings from the operational environment to the data warehouse which includes data extraction, cleaning and transformation rules.
5. Data related to system performance which include indices and profiles that improve data access and retrieval performance.

4.3   Data Warehouse Back-end Tools

There are many challenges in creating and maintaining a large data warehouse. Firstly, a good database schema must be designed to hold an integrated collection of data copied from multiple sources. Secondly, after the warehouse schema is designed, the warehouse must be populated and over time, it must be kept consistent with the source databases. Data is *extracted* from external sources, *cleaned* to minimize errors and *transformed* to create aggregates and summary tables. Data warehouse systems use backend tools and utilities to populate and refresh their data. These tools are called Extract, Transform and Load (ETL) tools. They include the following functionality:

- *Data Cleaning:* Real world data tends to be incomplete, noisy and inconsistent [5]. The ETL tools provide *data cleaning* routines to fill in missing values, remove noise from the data and correct inconsistencies in the data. Some data inconsistencies can be detected by using the functional dependencies among attributes to find values that contradict the functional constraints. The system will provide capability for users to add rules for data cleaning.

- *Data Integration:* The data mining/analysis task requires combining data from multiple sources into a coherent data store [6]. These sources may be multiple sources or flat files. There are a number of issues to consider during data integration. Schema integration can be quite tricky. How can real-world entities from multiple data sources be matched up? For example, how can we make sure that customer ID in one database and cust number in another database refers to the same entity? Our application will use metadata to help avoid errors during data integration. Redundancy is another important issue for data integration. An attribute is redundant if it can be derived from another table. For example, annual revenue for a company can be derived from the monthly revenue table for a company. One method of detecting redundancy is by using *correlation analysis*. A third important issue in data integration is the *detection* and *resolution* of data value conflicts. For example, for the same real world entity, attribute values from different sources may differ. For example, the weight attribute may be stored in the metric unit in one system and in British imperial unit on the other system.

- *Data Transformation:* Data coming from input sources can be transformed so that it is more appropriate for data analysis [7]. Some examples of transformations that are supported in our system are as follows

  - *Aggregation:* Apply certain summarization operations to incoming data. For example, the daily sales data can be aggregated to compute monthly and yearly total amounts.
  - *Generalization:* Data coming from input sources can be generalized into higher-level concepts through the use of concept hierarchies. For example, values for numeric attributes like age can be mapped to higher-level concepts such as *young, middle age, senior.*
  - *Normalization:* Data from input sources is scaled to fall within a specified range such as 0.0 to 1.0
  - *Data Reduction:* If the input data is very large complex data analysis and data mining can take a very long time making such analysis impractical or infeasible. Data reduction techniques can be used to reduce the data set so that analysis on the reduced set is more efficient and yet produce the same analytical results. The following are some of the techniques for data reduction that are supported in our system.
    a) Data Cube Aggregation: Aggregation operators are applied to the data for construction of data cubes.

b) Dimension Reduction: This is accomplished by detecting and removing irrelevant dimensions.

c) Data Compression: Use encoding mechanisms to reduce the data set size.

d) Concept Hierarchy Generation: Concept hierarchies allow mining of data at multiple levels of abstraction and they are a powerful tool for data mining.

- *Data Refreshing:* The application will have a scheduler that will allow the user to specify the frequency at which the data will be extracted from the source databases to refresh the data warehouse.

4.4   Views and Data Warehouse

Views are often used in data warehouse applications. OLAP queries are typically aggregate queries. Analysts often want fast answers to these queries over very large data sets and it is natural to consider pre-computing views and the aggregates. The choice of views to materialize is influenced by how many queries they can potentially speed up and the amount of space required to store the materialized view.

A popular approach to deal with the problem is to evaluate the view definition and store the results. When a query is now posed on the view, the query is executed directly on the pre-computed result. This approach is called *view materialization* and it results in fast response time. The disadvantage is that we must maintain consistency of the materialized view when the underlying tables are updated.

There are three main questions to consider with regard to view materialization.

1. What views to materialize and what indexes to create.
2. How to utilize the materialized view to answer a query
3. How often should the materialized view be refreshed.

5.   COMMERCIAL DATA WAREHOUSE TOOLS

The following is a summary of commercial data warehouse tools that are available in the market.

1. Back End ETL Tools

- DataStage: This was originally developed by Ardent Software and it is now part of Ascential Software. See http://www.ascentialsoftware.com
- Informatica is an ETL tool for data warehousing and it provides analytic software that for business intelligence. See http://www.informatica.com
- Oracle: Oracle has a set of data warehousing tools for OLAP and ETL functionality. See http://www.oracle.com
- DataJunction: See http://www.datajunction.com

2. Multidimensional Database Engines: Arbor ESSbase, SAS system
3. Query/OLAP Reporting Tools: Brio, Cognos/Impromptu, Business Objects, Mirostrategy/DSS, Crystal reports

6.   FROM DATA WAREHOUSING TO DATA MINING

In this section, we study the usage of data warehousing for data mining and knowledge discovery. Business executives use the data collected in a data warehouse for data analysis and make strategic business decisions. There are three kinds of applications for a data warehouse. Firstly, *Information Processing* supports querying, basic statistical analysis and reporting. Secondly, *Analytical Processing* supports multidimensional data analysis using slice-and-dice and drill-down operations. Thirdly, *Data Mining* supports knowledge discovery by finding hidden patterns and associations and presenting the results using visualization tools. The process of knowledge discovery is illustrated in the figure 4 and it consists of the following steps:

a) Data cleaning: removing invalid data
b) Data integration: combine data from multiple sources
c) Data transformation: data is transformed using summary or aggregation operations
d) Data mining: apply intelligent methods to extract patterns
e) Evaluation and presentation: use visualization techniques to present the knowledge to the user

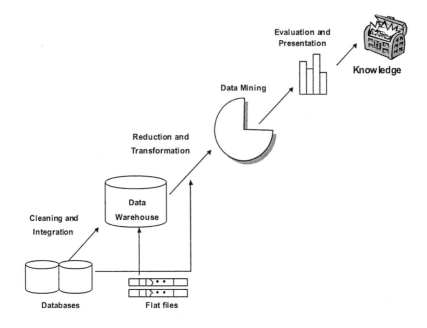

Figure 4: Architecture of the Knowledge Discovery Process

6.1   Data Mining Techniques

The following are different kinds of techniques and algorithms that data mining can provide.

a) Association Analysis: This involves discovery of *association rules* showing attribute-value conditions that occur frequently together in a given set of data. This is used frequently for market basket or transaction data analysis. For example, the following rule says that if a customer is in age group 20 to 29 years and income is greater than 40K/year then he or she is likely to buy a DVD player.

Age(X, "20-29") & income(X, ">40K") => buys (X, "DVD player")
[support = 2% , confidence = 60%]

Rule *support*   and *confidence* are two measures of rule interestingness. A support of 2% means that 2% of all transactions under analysis show that this rule is true. A confidence of 60% means that among all customers in the age group 20-29 and income greater than 40K, 60% of them bought DVD players.

A popular algorithm for discovering association rules is the Apriori method. This algorithm uses an iterative approach known as *level-wise* search where k-itemsets are used to explore (k+1) itemsets. Association rules are widely used for prediction.

b) Classification and Prediction: Classification and prediction are two forms of data analysis that can be used to extract models describing important data classes or to predict future data trends. For example, a classification model can be built to categorize bank loan applications as either safe or risky. A prediction model can be built to predict the expenditures of potential customers on computer equipment given their income and occupation. Some of the basic techniques for data classification are decision tree induction, Bayesian classification and neural networks.

These techniques find a set of models that describe the different *classes* of objects. These models can be used to predict the class of an object for which the class is unknown. The derived model can be represented as rules (IF-THEN), decision trees or other formulae.

c) Clustering: This involves grouping objects so that objects within a cluster have high similarity but are very dissimilar to objects in other clusters. Clustering is based on the principle of *maximizing the intraclass similarity and minimizing the interclass similarity.*

In business, clustering can be used to identify customer groups based on their purchasing patterns. It can also be used to help classify documents on the web for information discovery. Due to the large amount of data collected, cluster analysis has recently become a highly active topic in data mining research. As a branch of statistics, cluster analysis has been extensively studied for many years, focusing primarily on *distance based cluster analysis.* These techniques have been built into statistical analysis packages such as S-PLUS and SAS. In machine learning, clustering is an example of *unsupervised learning.* For this reason clustering is an example of *learning by observation.*

d) Outlier Analysis: A database may contain data objects that do not comply with the general   model or behavior of data. These data objects are called *outliers.* These outliers are useful for applications such as fraud detection and network intrusion detection.

6.2   Research Issues in Data Mining

In this section, we briefly discuss some of the research issues in data mining.

a) Mining methodology and user interaction issues:
- Data mining query languages
- Presentation and visualization of data mining results
- Data cleaning and handling of noisy data

b) Performance Issues:
- Efficiency and scalability of data mining algorithms
- Coupling with database systems
- Parallel, distributed and incremental mining algorithms
- Handling of complex data types such as multimedia, spatial data and temporal data

6.3   Applications of Data Mining

Data mining is expected to have broader applications as compared to OLAP. It can help business managers find and reach suitable customers as well as develop special intelligence to improve market share and profits. Here are some applications of data mining.

1. DNA Data Analysis: A great deal of biomedical research is focused on DNA data analysis. Recent research in DNA data analysis has enabled the discovery of genetic causes of many diseases as well as discovery of new medicines. One of the important search problems in genetic analysis is similarity search and comparison among the DNA sequences. Data mining techniques can be used to solve these problems.
2. Intrusion Detection and Network Security: This will be discussed further in later chapters.
3. Financial Data Analysis: Most financial institutions offer a variety of banking services such as credit and investment services. Data warehousing techniques can be used to gather the data to generate monthly reports. Data mining techniques can be used to predict loan payments and customer credit policy analysis.
4. Data Analysis for Retail Industry: Retail is a big application of data mining since it collects huge amount of data on sales, shopping history and service records. Data mining techniques can be used for

multidimensional analysis of sales, and customers by region and time. It can also be used to analyze effectiveness of sales campaigns.

5. Data Analysis for Telecom Industry: The following are some examples of where data mining can be used to improve telecom services:
   - Analysis of calling patterns to determine what kind of calling plans to offer to improve profitability.
   - Fraud detection by discovering unusual patterns
   - Visualization tools for data analysis.

6.4   Commercial Tools for Data Mining

In this section, we briefly outline a few typical data mining systems in order to give the reader an idea about what can be done with the current data mining products.

- **Intelligent Miner** is an IBM data mining product that provides a wide range of data mining algorithms including association, classification, predictive modeling and clustering. It also provides an application toolkit for neural network algorithms and data visualization. It includes scalability of mining algorithms and tight integration with IBM's DB2 relational database systems.
- **Enterprise Miner** was developed by SAS Institute, Inc. It provides multiple data mining algorithms including regression, classification and statistical analysis packages. One of it's distinctive feature is the variety of statistical analysis tools, which are built based on the long history of SAS in the market for statistical analysis.
- **MineSet** was developed by Silicon Graphics Inc. (SGI). It also provides multiple data mining algorithms and advanced visualization tools. One distinguishing feature of *MineSet* is the set of robust graphics tools such as rule visualizer, tree visualizer and so on.
- **Clementine** was developed by Integral Solutions Ltd. (ISL). It provides an integrated data mining development environment for end users and developers. It's object oriented extended module interface allows user's algorithms and utilities to be added to Clementine's visual programming environment.
- **DBMiner** was developed by DBMiner Technology Inc. It provides multiple data mining algorithms including discovery driven OLAP analysis, association, classification and clustering. A distinct feature of *DBMiner* is its data cube based analytical mining.

There are many other commercial data mining products, systems and research prototypes that are also fast evolving. Interested readers can consult surveys on data warehousing and data mining products.

7.   DATA ANALYSIS APPLICATIONS FOR NETWORK/WEB SERVICES

In this section we discuss our experience [8] [9] [10], [11] [12] in developing data analysis applications using data warehousing, OLAP and data mining technology for AT&T Business Services. AT&T Business Services (ABS) designs, manages and operates global networks for multinational corporations. Global Enterprise Management System (GEMS) is a platform that is used by ABS to support design, provisioning and maintenance of the network (LANs, WANS, intranets etc.) and desktop devices for multinational corporations such as BANCONE and CITICORP. The primary functions supported by GEMS are: ticketing, proactive management of client's networks, client's asset management, network engineering and billing. GEMS applications use an Integrated Database to store fault tickets, assets and inventory management information.

The main purpose of GEMS DW is for ABS to generate reports about the performance and reliability of the network and compare it with the system level agreements (SLAs) that ABS has agreed to provide to its client companies. An SLA is a contract between the service provider and a customer (usually an enterprise) on the level of service quality that should be delivered. An SLA can contain the following metrics:
1. Mean Time To Repair (MTTR) a fault
2. Available network bandwidth (e.g. 1.3 Mbps, 90% of the time on 80% of user nodes)
3. Penalty (e.g. $10,000) if agreement is not met.

SLAs give service providers a competitive edge for selling network/web services into the consumer market and maintain customer satisfaction. In order to track SLAs, service providers have to generate user reports on satisfaction/violation of the metrics. In addition, the provider must have the ability to drill down to detailed data in response to customer inquires.

The DW enables the knowledge worker (executive, manager, and analyst) to track the SLAs. For example, the DW is used to generate monthly reports for a client and to gather statistics such as Mean Time to Repair

(MTTR) and average number of fault tickets that are open for an ABS client company. The main reason to separate the decision support data from the operation data is performance.   Operational databases are designed for known transaction workloads.   Complex queries/reports degrade the performance of the operational databases.   Moreover special data organization and access methods are required for optimizing the report generation process. This project also required data integration and data fusion from many external sources such as operational databases and flat files.

The main components used in our system are as follows.

1. Ascential's DataStage Tool is an Extraction-Transformation-Load-Management (ETLM) class of tool that defines how data is *extracted* from a data source, *transformed* by the application of functions, joins and possibly external routines, and then *loaded* into a target data source.
2. DataStage reads data from the source information repositories and it applies transformations as it loads all data into a repository (atomic) database.
3. Once the atomic data repository is loaded with all source information a second level of ETL transformations is applied to various data streams to create one or more Data Marts. Data Marts are a special sub-component of a data warehouse in that they are highly de-normalized to support the fast execution of reports. Some of these Data Marts are created using Star Schemas.
4. Both the atomic data repository and the data marts are implemented using Oracle version 8i DBMS.
5. Once the atomic repository and the data marts have been populated , OLAP tools such as COGNOS and ORACLE EXPRESS are configured to access both the data marts as well as the atomic repository in order to generate the monthly reports.

An architecture diagram of our system is shown in Figure 5

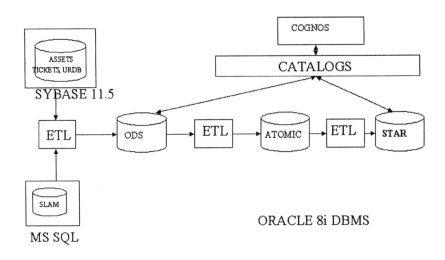

Figure 5: Architecture of the GEMS Data Warehouse System

The main advantages our system are:

1. Since a separate DW system is used to generate the reports the time taken to generate the reports is much better. Also, the execution of reports does not impact the applications that are using the source databases.
2. The schemas in the DW are optimized by using de-normalization and pre-aggregation techniques. This results in much better execution time for reports.

Some of the open research problems that we are currently investigating are:

- Time to refresh the data in the data warehouse was large and report generation activity had to be suspended until the time when changes were propagated into the DW. Therefore, there was a need to investigate incremental techniques for propagating the updates from source databases
- Loading the data in the data warehouse took a long time (10 to 15 hours). In case of any crashes, the entire loading process had to be re-started.

This further increased the down time for the DW and there was a need deal with crash recovery more efficiently.

• There was no good support for tracing the data in the DW back to the source information repositories.

7.1 Open Research Problems in Data Warehouse Maintenance

A critical part of data analysis systems is a component that can efficiently extract data from multiple sources, filter it to remove noise, transform it and then load it into the target data analysis platform. This process, which is used to design, deploy and manage the data marts is called the ETL (Extract, Transform and Load) process. There are a number of open research problems in designing the ETL process.

1. **Maintenance of Data Consistency:** Since source data repositories continuously evolve by modifying their content or changing their schema one of the research problems is how to incrementally propagate these changes to the central data warehouse. Both re-computation and incremental view maintenance are well understood for centralized relational databases. However, more complex algorithms are required when updates originate from multiple sources and affect multiple views in the Data Warehouse. The problem is further complicated if the source databases are going through schema evolution.

2. **Maintenance of Summary Tables:** Decision support functions in a data warehouse involve complex queries. It is not feasible to execute these queries by scanning the entire data. Therefore, a data warehouse builds a large number of *summary tables* to improve performance. As changes occur in the source databases, all summary tables in the data warehouse need to be updated. A critical problem in data warehouse is how to update these summary tables *efficiently and incrementally.*

3. **Incremental Resuming of Failed Loading Process:** Warehouse creation and maintenance loads typically take hours to run. Our experience in loading a data warehouse for network management applications at AT&T took about 10 to 15 hours. If the load is interrupted by failures, traditional recovery methods undo the changes. The administrator must then restart the load and hope that it does not fail again. More research is required into algorithms for resumption of the incomplete load so as to reduce the total load time.

4. **Tracing the Lineage of Data:** Given data items in the data warehouse, analysts often want to identify the source items and source databases that

produced those data items. Research is required for algorithms to trace the lineage of an item from a view back to the source data items in the multiple sources.

5. **Data Reduction Techniques:** If the input data is very large, data analysis can take a very long time making such analysis impractical or infeasible. There is a need for data reduction techniques that can be used to reduce the data set so that analysis on the reduced set is more efficient and yet produce the same analytical results.The following are examples of some of the algorithmic techniques that can be used for data reduction.

- Data Cube Aggregation: Aggregation operations such as AVERAGE(), SUM() and COUNT() can be applied to input data for construction of data cubes. These operations reduce the amount of data in the DW and also improve the execution time for decision support queries on data in the DW
- Dimension Reduction: This is accomplished by detecting and removing irrelevant attributes that are not required for data analysis. Data Compression: Use encoding mechanisms to reduce the data set size.
- Concept Hierarchy Generation: Concept hierarchies allow analysis of data at multiple levels of abstraction and they are a powerful tool for data analysis. For example, values for numeric attributes like age can be mapped to higher-level concepts such as *young, middle age, senior.*

6. **Data Integration and Data Cleaning Techniques:** Generally, data analysis task includes *data integration,* which combine data from multiple sources into a coherent data store. These sources may include multiple databases or flat files. A number of problems can arise during data integration. Real world entities in multiple data sources can be given different names. How does an analyst know that *employee-id* in one database is same as *employee-number* in another database. We plan to use meta-data to solve the problem of data integration. Data coming from input sources tends to be incomplete, noisy and inconsistent. If such data is directly loaded in the DW it can cause errors during the analysis phase resulting in incorrect results. *Data cleaning* methods will attempt to smooth out the noise, while identifying outliers, and correct inconsistencies in the data. We are investigating the following techniques for noise reduction and data smoothing.

a) **Binning:** These methods smooth a sorted data value by consulting the values around it.

b) **Clustering:** Outliers may be detected by clustering, where similar values are organized into groups or clusters. Intuitively, values that fall outside of the set of clusters may be considered outliers.

c) **Regression:** Data can be smoothed by fitting the data to a function, such as with regression. Using regression to find a mathematical equation to fit the data helps smooth out the noise.

Data pre-processing is an important step for data analysis. Detecting data integration problems, rectifying them and reducing the amount of data to be analyzed can result in great benefits during the data analysis phase.

7.2   Current Research in the area of Data Warehouse Maintenance

A number of techniques for view maintenance and propagation of changes from the source databases to the data warehouse (DW) have been discussed in literature. [5] [14] describes techniques for view maintenance and refreshing the data in a DW.

[15] also describes techniques for maintenance of data cubes and summary tables in a DW environment. However, the problem of propagating changes in a DW environment is more complicated due to the following reasons:

a) In a DW, data is not refreshed after every modification to the base data. Rather, large batch updates to the base data must be considered which requires new algorithms and techniques.

b) In a DW environment, it is necessary to transform the data before it is deposited into the DW. These transformations may include aggregating or summarizing the data.

c) The requirements of data sources may change during the life cycle, which may force schema changes for the data source. Therefore techniques are required that can deal with both source data changes and schema changes. [Liu 2002] describes some techniques for dealing with schema changes in the data sources.

[6], [13] describes techniques for practical lineage tracing of data in a DW environment. It enables users to "drill through" from the views in the DW all the way to the source data that was used to create the data in the DW. However, their methods lack techniques to deal with historical source data or data from previous source versions.

8.   CONCLUSIONS

A data warehouse is a subject oriented collection of data that is used for decision support systems. They typically use a multidimensional data model to facilitate data analysis. They are implemented using a three tier architecture. The bottom most tier is a database server which is typically a RDBMS. The middle tier is a OLAP server and the top tier is a client, containing query and reporting tools. Data mining is the task of discovering interesting patterns from large amounts of data where data can be stored in multiple repositories. Efficient data warehousing and data mining techniques are challenging to design and implement for large data sets.

In this chapter, we have given a summary of Data Warehousing, OLAP and Data Mining Technology. We have also described our experience in using this technology to build Data Analysis Application for Network/Web services. We have also described some open research problems that need to be solved in order to efficiently extract data from distributed information repositories. Although, some commercial tools are available in the market, our experience in building a decision support system for a network/web services has shown that they are inadequate. We believe that there are several important research problems that need to be solved to build flexible, powerful and efficient data analysis applications using data warehousing and data mining techniques.

References

1.   S. Chaudhuri, U. Dayal: An Overview of Data Warehousing and OLAP Technology, SIGMOD Record, March 1997.
2.   W.H. Inmon: Building the Data Warehouse (2nd Edition) John Wiley, 1996.
3.   R. Kimball: The Data Warehouse Toolkit, John Wiley, 1996.
4.   D. Pyne: Data Preparation for Data Mining, San Francisco, Morgan Kaufmann, 1999
5.   Prabhu Ram and Lyman Do: Extracting Delta for Incremental Warehouse, Proceedings of IEEE 16th Int. Conference on Data Engineering, 2000.
6.   Y. Cui and J. Widom: Practical Lineage Tracing in Data Warehouses, Proceedings of IEEE 16th Int. Conference on Data Engineering, 2000.
7.   S. Chaudhuri, G. Das and V. Narasayya: A Robust, Optimization Based Approach for Approximate Answering of Aggregate Queries, Proceeding of ACM SIGMOD Conference, 2001, pp 295-306
8.   Anoop Singhal, "Design of a Data Warehouse for Network/Web Services", "Proceedings of Conference on Information and Knowledge Management, CIKM 2004.
9.   Anoop Singhal, "Design of GEMS Data Warehouse for AT&T Business Services", Proceedings of AT&T Software Architecture Symposium, Somerset, NJ, March 2000
10.   ANSWER: Network Monitoring using Object Oriented Rules" (with G. Weiss and J. Ros), Proceedings of the Tenth Conference on Innovative Application of Artificial Intelligence, Madison, Wisconsin, July 1998.

11. "A Model Based Approach to Network Monitoring", Proceedings of ACM Workshop on Databases: Active and Real Time, Rockville, Maryland Nov. '96 pages 41-45.

12. Jiawei Han, Micheline Kamber, Data Mining: Concepts and Techniques, Morgan Kaufmann, August 2000

13. Jennifer Widom, "Research Problems in Data warehousing", Proc. Of 4th Int'l Conference on Information and Knowledge Management, Nov. 1995

14. Hector Garcia Molina, "Distributed and Parallel Computing Issues in Data Warehousing", Proc. Of ACM Conference on Distributed Computing, 1999.

15. A. Gupta and I.S. Mumick, "Maintenance of Materialized Views", IEEE Data Engineering Bulletin, June 1995

16. Vipin Kumar et al., Data Mining for Scientific and Engineering Applications, Kluwer Publishing 2000

17. Bernstein P., Principles of Transaction Processing, Morgan Kaufman, San Mateo, CA 1997

18. Miller H and Han J, Geographic Data Ming and Knowledge Discovery, UK 2001

19. Liu, Bin, Chen, Songting and Rundensteiner, E. A. Batch Data Warehouse Maintenance in Dynamic Environment .   In Proceedings of CIKM' 2002, McLean, VA, Nov. 2002

20. Hector Garcia Molina, J.D. Ullman, J. Widom, Database Systems the Complete Book, Prentice Hall, 2002

Chapter 2

NETWORK AND SYSTEM SECURITY

Anoop Singhal

Abstract: This chapter discusses the elements of computer security such as
              authorization, authentication and integrity. It presents threats against
              networked applications such as denial of service attacks and protocol attacks.
              It also presents a brief discussion on firewalls and intrusion detection systems

Key words: computer virus, worms, DOS attacks, firewall, intrusion detection

Computer security is of importance to a wide variety of practical domains ranging from banking industry to multinational corporations, from space exploration to the intelligence community and so on. The following principles are the foundation of a good security solution:

- Authentication: The process of establishing the validity of a claimed identity.
- Authorization: The process of determining whether a validated entity is allowed access to a resource based on attributes, predicates, or context.
- Integrity: The prevention of modification or destruction of an asset by an unauthorized user.
- Availability: The protection of assets from denial-of-service threats that might impact system availability.
- Confidentiality: The property of non-disclosure of information to unauthorized users.
- Auditing: The property of logging all system activities

Computer security attempts to ensure the confidentiality, integrity and availability of computing resources. The principal components of a computer that need to be protected are hardware, software and the communication links. This chapter describes different kind of threats related to computer security and protection mechanisms that have been developed to protect the different components.

1.   VIRUSES AND RELATED THREATS

This section briefly discusses a variety of software threats. We first present information about computer viruses and worms followed by techniques to handle them.

A virus is a program that can "infect" other programs by modifying them and inserting a copy of itself into the program. This copy can then go to infect other programs. Just like its biological counterpart, a computer virus carries in its instructional code the recipe for making perfect copies of itself. A virus attaches itself to another program and then executes secretly when the host program is run.

During it lifetime a typical virus goes through the following stages:

Dormant Phase: In this state the virus is idle waiting for some event to happen before it gets activated. Some examples of these events are date/timestamp, presence of another file or disk usage reaching some capacity.

Propagation Phase: In this stage the virus makes an identical copy of itself and attaches itself to another program. This infected program contains the virus and will in turn enter into a propagation phase to transmit the virus to other programs.

Triggering Phase: In this phase the virus starts performing the function it was intended for. The triggering phase can also be caused by a set of events.

Execution Phase: In this phase the virus performs its function such as damaging programs and data files.

1.1   Types of Viruses

The following categories give the most significant types of viruses.

Parasitic Virus: This is the most common kind of virus. It attaches itself to executable files and replicates when that program is executed.
Memory Resident Virus: This kind of virus resides in main memory. When ever a program is loaded into memory for execution, it attaches itself to that program.
Boot Sector Virus: This kind of virus infects the boot sector and it spreads when the system is booted from the disk.
Stealth Virus: This is a special kind of virus that is designed to evade itself from detection by antivirus software.
Polymorphic virus: This kind of virus that mutates itself as it spreads from one program to the next, making it difficult to detect using the "signature" methods.

1.2   Macro Viruses

In recent years macro viruses have become quite popular. These viruses exploit certain features found in Microsoft Office Applications such as MS Word or MS Excel. These applications have a feature called *macro* that people use to automate repetitive tasks. The macro is written in a programming language such as Basic. The macro can be set up so that it is invoked when a certain function key is pressed. Certain kinds of macros are auto execute, they are automatically executed upon some events such as starting the execution of a program or opening of a file. These auto execution macros are often used to spread the virus. New version of MS Word provides mechanisms to protect itself from macro virus. One example of this tool is a Macro Virus Protection tool that can detect suspicious Word files and alert the customer about a potential risk of opening a file with macros.

1.3   E-mail Viruses

This is a new kind of virus that arrives via email and it uses the email features to propagate itself. The virus propagates itself as soon as it is activated (typically by opening the attachment) and sending an email with the attachment to all e-mail addresses known to this host. As a result these viruses can spread in a few hours and it becomes very hard for anti-virus software to respond before damage is done.

1.4   Worms

A virus typically requires some human intervention (such as opening a file) to propagate itself. A worm on the other hand typically propagates by itself. A worm uses network connections to propagate from one machine to another. Some examples of these connections are:

Electronic mail facility
Remote execution facility
Remote login facility

A worm will typically have similar phases as a virus such as dormant phase, a propagation phase, a triggering phase and an execution phase. The propagation phase for a worm uses the following steps:

Search the host tables to determine other systems that can be infected.
Establish a connection with the remote system
Copy the worm to the remote system and cause it to execute

Just like virus, network worms are also difficult to detect. However, properly designed system security applications can minimize the threat of worms.

1.5   The Morris Worm

This worm was released into the internet by Robert Morris in 1998. It was designed to spread on UNIX systems and it used a number of techniques to propagate. In the beginning of the execution, the worm would discover other hosts that are known to the current host. The worm performed this task by examining various list and tables such as machines that are trusted by this host or user's mail forwarding files. For each discovered host, the worm would try a number of methods to login to the remote host:

Attempt to log on to a remote host as a legitimate user.
Use the finger protocol to report on the whereabouts of a remote user.
Exploit the trapdoor of a remote process that sends and receives email.

1.6   Recent Worm Attacks

One example of a recent worm attack is the Code Red Worm that started in July 2001. It exploited a security hole in the Microsoft Internet

Information Server (IIS) to penetrate and spread itself. The worm probes random IP addresses to spread to other hosts. Also during certain periods of times it issues denial of service attacks against certain web sites by flooding the site with packets from several hosts. Code Red I infected nearly 360,000 servers in 14 hours. Code Red II was a second variant that targeted Microsoft IIS.

In late 2001, another worm called Nimda appeared. The worm spread itself using different mechanisms such as
Client to client via email
From web server to client via browsing of web sites
From client to Web server via exploitation of Microsoft IIS vulnerabilities

The worm modifies Web documents and certain executables files on the infected system.

1.7   Virus Counter Measures

Early viruses were relatively simple code fragments and they could be detected and purged with simple antivirus software. As the viruses got more sophisticated the antivirus software packages have got more complex to detect them.

There are four generations of antivirus software:

First Generation: This kind of scanner requires a specific signature to identify a virus. They can only detect known viruses.
Second Generation: This kind of scanner does not rely on a specific signature. Rather, the scanner uses heuristic rules to search for probable virus infections. Another second generation approach to virus detection is to use integrity checking. For example, a checksum can be appended to every program. If a virus infects the program without changing the checksum, then an integrity check will detect the virus.
Third Generation: These kind of programs are memory resident and they identify a virus by its actions rather than by its structure. The advantage of this approach is that it is not necessary to generate signature or heuristics. This method works by identifying a set of actions that indicate some malicious work is being performed and then to intervene.

Fourth Generation: These kind of packages consist of a variety of antivirus techniques that are used in conjunction. They including scanning, access control capability which limits the ability of a virus to penetrate the system and update the files to propagate the infection.

2.   PRINCIPLES OF NETWORK SECURITY

In the modern world we interact with networks on a daily basis such as when we perform banking transactions, make telephone calls or ride trains and planes. Life without networks would be considerably less convenient and many activities would be impossible. In this chapter, we describe the basics of computer networks and how the concepts of confidentiality, integrity and availability can be applied for networks.

2.1   Types of Networks and Topologies

A network is a collection of communicating hosts. There are several types of networks and they can be connected in different ways. This section provides information on different classes of networks.

a) Local Area Networks: A local area network (or LAN) covers a small distance, typically within a single building. Usually a LAN connects several computers, printers and storage devices. The primary advantage of a LAN to users is that it provides shared access to resources such programs and devices such as printers.

b) Wide Area Networks: A wide are network differs from a local area network in terms of both size and distance. It typically covers a wide geographical area. The hosts on a WAN may belong to a company with many offices in different cities or they may be a cluster of independent organizations within a few miles of each other who would like to share the cost of networking.   Therefore a WAN could be controlled by one organization or it can be controlled by multiple organizations.

c) Internetworks (Internets): Network of networks or internet is a connection of two or more separate networks in that they are separately managed and controlled. The Internet is a collection of networks that is loosely controlled by the Internet Society. The Internet Society enforces certain minimal rules to make sure that all users are treated fairly.

2.2   Network Topologies

The security of a network is dependent on its topology. The three different topologies are as follows.

a)   Common Bus: Conceptually, a common bus is a single wire to which each node of a LAN is connected. In a common bus, the information is broadcast and nodes must continually monitor the bus to get the information addressed to it.

b)   Star or Hub: In this topology each node is connected to a central "traffic controller" node. All communication flows from the source node to the traffic controller node and from the traffic controller node to the other nodes.

c)   Ring: In this architecture, each node receives many messages, scans each and removes the one designated for itself. In this topology, there is no central node. However, there is one drawback with this architecture. If a node fails to pass a message that it has received, the other nodes will not be able to receive that information.

3.   THREATS IN NETWORKS

Network security has become important due to the inter-connection of computers and the rise of the internet. This section describes some of the popular network threats.

a) **Spoofing:** By obtaining the network authentication credentials of an entity (such as a user or a process) permits an attacker to create a full communication under the entity's identity. Examples of spoofing are masquerading and man-in-the-middle attack.

b) **Masquerade:** In a masquerade a user who is not authorized to use a computer pretends to be a legitimate user. A common example is URL confusion. Thus abc.com, abc.org or abc.net might be three different organizations or one legitimate organization and two masquerade attempts from some one who registered similar names.

c) **Phishing Attacks:** These attacks are becoming quite popular due to the proliferation of Web sites. In *phishing scams,* an attacker sets up a web site that masquerades as a legitimate site. By tricking a user, the phishing site obtains the user's cleartext password for the legitimate site. Phishing has proven to be quite effective in stealing user passwords.

d) **Session Hijacking:**   It is intercepting and carrying out a session begun by another entity. Suppose two people have entered into a session but then a third person intercepts the traffic and carries out a session in the name of the other person then this will be called session hijacking. For example, if an Online merchant used a wiretap to intercept packets between you and Amazon.com, the Online merchant can monitor the flow of packets. When the user has completed the order, Online merchant can intercept when the "Ready to check out" packet is sent and finishes the order with the user obtaining shipping address, credit card detail and other information. In this case we say the Online merchant has hijacked the session.

e) **Man-in-the-Middle Attack:** In this type of attack also one entity intrudes between two others. The difference between man-in-the-middle and hijacking is that a man-in-the-middle usually participates from the start of the session, whereas a session hijacking occurs after a session has been established. This kind of attack is frequently described in protocols. For example, suppose two parties want to exchange encrypted information. One party contacts the key server to get a secret key that will be used in the communication. The key server responds by sending the private key to both the parties. A malicious middleman intercepts the response key and then eavesdrop on the communication between the two parties.

f) **Web Site Defacement:** One of the most widely known attacks is the web site defacement attack. Since this can have a wide impact they are often reported in the popular press. Web sites are designed so that their code can be easily downloaded enabling an attacker to obtain the full hypertext document. One of the popular attacks against a web site is *buffer overflow.* In this kind of attack the attacker feeds a program more data than what is expected. A buffer size is exceeded and the excess data spills over adjoining code and data locations.

g) Message Confidentiality Threats:

- **Misdelivery:** Sometimes messages are misdelivered because of some flaw in the network hardware or software. We need to design mechanisms to prevent this.
- **Exposure:** To protect the confidentiality of a message, we must track it all the way from its creation to its disposal.
- **Traffic Flow Analysis:** Consider the case during wartime, if the enemy sees a large amount of traffic between the headquarters and a particular unit, the enemy will be able to infer that a significant action is being planned at that unit. In these situations there is a need to protect the contents of the message as well as how the messages are flowing in the network.

4.   DENIAL OF SERVICE ATTACKS

So far we have presented attacks that lead to failures of confidentiality or integrity. Availability attacks in network context are called denial of service attacks and they can cause a significant impact. The following are some sample denial of service attacks.

Connection Flooding: This is the most primitive denial-of-service attack. If an attacker sends so much data that the communication system cannot handle it then you are prevented from receiving any other data.

Ping of Death: Since ping requires the recipient to respond to the ping request, all that the recipient needs to do it to send a flood of pings to the intended victim.

Smurf: This is a variation of a ping attack. It uses the same vehicle, a ping packet with two extra twists. First, the attacker chooses a network of victims. The attacker spoofs the source address in the ping packet so that it appears to come from the victim. Then, the attacker sends this request to the network in broadcast mode by setting the last byte of the address to all 1s; broadcast mode packets are distributed to all the hosts.

Syn Flood: The attacker can deny service to the target by sending many SYN requests and never responding with ACKs. This fills up the victim's SYN_RECV queue. Typically, the SYN_RECV queue is quite small (about 10 to 20 entries). Attackers using this approach do one more thing, they spoof the nonexistent return address in the initial SYN packet.

4.1   Distributed Denial of Service Attacks

In order to perpetrate a distributed denial of service attack, an attacker does two things. In the first step, the attacker uses any convenient step (such as exploiting a buffer overflow) to plant a Trojan horse on a target machine. The installation of the Trojan horse as a file or a process does not attract any attention. The attacker repeats this process with many targets. Each of these targets then become what is known as a *zombie*. The target system carry out their work , unaware of the resident zombie.

At some point, the attacker chooses a victim and sends a signal to all the zombies to launch the attack. Then, instead of the victim trying to defend against one denial-of-service attack from one malicious host, the victim must try to counter n attacks from n zombies all acting at one.

4.2   Denial of Service Defense Mechanisms

The increased frequency of Denial of Service attacks has led to the development of numerous defense mechanisms. This section gives a summary of the taxonomy of defense mechanisms based on this paper.

Classification by Activity Level

Based on the activity level defense mechanisms can be classified into *preventive* and *reactive* mechanisms.

Preventive Mechanisms
The goal of these mechanisms is to either eliminate the possibility of DOS attacks or to endure the attack without denying services to legitimate clients.

Attack Prevention Mechanisms
These mechanisms modify the system configuration to eliminate the possibility of a DOS attack. System security mechanisms increase the overall security by guarding against illegitimate access from other machines. Examples of system security mechanisms include monitored access to the machine, install security patches, and firewall systems.

Protocol security mechanisms address the problem of bad protocol design which can be misused to exhaust the resources of a server by initiating a large number of such transactions. Classic misuse examples are the TCP

SYN attacks and the fragmented packet attack. An example of a protocol security mechanism is to have a design in which resources are committed to the client only after sufficient authentication is done.

Reactive Mechanisms

Reactive mechanisms alleviate the impact of an attack by *detecting* an attack and *responding* to it. Reactive mechanisms can be classified based on the mechanisms that they use *pattern detection, anomaly detection and hybrid detection.*

Mechanism with Pattern Attack Detection

In this method, signatures of known attacks are stored in a database. Each communication is monitored and compared with the database entries to discover the occurrence of an attack. Occasionally, the database is updated with new attack signatures. The obvious drawback of this detection mechanism is that it can only detect known attacks. On the other hand the main advantage is that known attacks are reliably detected and no false positives are encountered.

Mechanism with Anomaly Attack Detection

Mechanisms that deploy anomaly detection have a model of normal system behavior such as traffic or system performance. The current state of the system is periodically compared with the models to detect anomalies. The advantage of these techniques as compared to pattern detection is that unknown attacks can be discovered. However, they have to solve the following problems

Threshold setting: Anomalies are detected based on known settings. The setting of a low threshold leads to many false positives, while a high threshold reduces the sensitivity of the detection mechanism.

Model Update: Systems and communication patterns evolve with time and models need to be updated to reflect this change.

Mechanisms with Hybrid Attack Detection

These techniques combine the pattern based and anomaly-based detection, using data about attacks discovered through an anomaly detection mechanism to devise new attack signatures and update the database. Many intrusion detection systems use this technique but they have to be carefully designed.

Attack Response

The goal of the attack response is to mitigate the impact of attack on a victim machine so as to minimize the collateral damage to clients of the victim. Reactive mechanisms can be classified based on the response strategy into *agent identification, filtering and reconfiguration* approaches.

Agent Identification Mechanisms

These mechanisms provide the victim with information about the identity of the machines that are responsible to perform the attacks. This information can be combined with other response approaches to reduce the impact of attacks.

Filtering Mechanism

These techniques use the information provided by a detection mechanism to filter out the attack stream completely. A dynamically deployed firewall is an example of such a system.

Reconfiguration System

These mechanisms change the connectivity of the victim or the intermediate network topology to isolate the attack machines. One example of such a system is a reconfigurable overlay network.

5.   NETWORK SECURITY CONTROLS

Encryption

Encryption is the most important and versatile tool for network security experts. It can provide privacy, authenticity, integrity and limited access to data. Encryption can be applied wither between two hosts (link encryption) or between two applications (called end-to-end encryption).

Link Encryption

Link encryption protects the message in transit between two computers, however the message is in clear text inside the host. In this method, the data is encrypted before it is placed on the physical communication link. The encryption occurs at the lowest layer 1 or 2 in the OSI model. Similarly, decryption occurs when the data arrives at the receiving computer. This mechanism is really useful when the transmission point is of greatest vulnerability.

End-to-end Encryption

This mechanism provides security from one end of transmission to the other. In this case encryption is performed at the highest levels (layer 7 or layer 6).

Virtual Private Networks

Link encryption can be used to give the same protection to a user as of they are on a private network, even when their communication links are part of a public network.

Firewalls can be used to implement a Virtual Private Network (VPN). When a user first requests communication with a firewall, the user can request a VPN session with the firewall. The user and the firewall can agree on a session encryption key and the user can use that key for all subsequent communication. With a VPN all communication passes through an *encrypted tunnel.*

PKI and Certificates

A **public key infrastructure (PKI)** is a process created to enable users to implement public key cryptography usually in a distributed environment. PKI usually offers the following services

Create certificates that associates a user's identity to s cryptographic key
Distribute certificates from its database
Sign certificates to provide authenticity
Confirm a certificate if it is valid

PKI is really a set of policies, products and procedures with some flexibility for interpretation. The policies define a set of rules under which the system operates, it defines procedures on how to handle keys and how to manage the risks.

SSH Encryption

SSH is a protocol that is available under Unix and Windows 2000 that provides an authenticated and encrypted path to the shell or the OS command interpreter. SSH protects against spoofing attacks and modification of during in communication.

SSL Encryption

The Secure Sockets Layer (SSL) protocol was originally designed by Netscape to protect communication between a web browser and a server. It is also known as transport layer security (TLS). SSL interfaces between the applications (e.g. a browser) and the TCP/IP protocols to provide server authentication, client authentication and an encrypted communications channel between the client and server.

IPSec

The address space for Internet is running out as more machines and domain names are being added to the Internet. A new structure called **IPv6** solves this problem by providing a 64 bit address space to IP addresses. As part of IPv6, the Internet Engineering Task Force (IETF) adopted an **IP Security Protocol (IPSec) Suite** that addresses problems such as spoofing, eavesdropping and session hijacking. IPSec is implemented at the IP layer so it affects all layers above it. IPSec is somewhat similar to SSL, in that it supports authentication and confidentiality that does not necessitate significant changes either above it (in applications) or below it (in the TCP protocols). Just like SSL, it was designed to be independent of the cryptographic protocols and to allow the two communicating parties to agree on a mutually supported set of protocols.

The basis of IPSec is called a security association which is basically a set of security parameters that are required to establish a secured communication. Some examples of these parameters are:

Encryption algorithm and mode
Encryption Key
Authentication protocol and key
Lifespan of the association to permit long running sessions to select a new key
Address of the opposite end of an association

6.   FIREWALLS

6.1   What they are

A firewall is a device that filters all traffic between a "protected" network and the "outside" network. Generally, a firewall runs on a dedicated machine which is a single point through which all the traffic is channeled. The purpose of a firewall is to keep "malicious" things outside a protected environment. For example, a firewall may impose a policy that will permit traffic coming from only certain IP addresses or users.

6.2   How do they work

There are different kind of firewalls.

Packet Filtering Firewall

It is the simplest form of firewall and in some situations it is most effective. It is based on certain packet address (source or destination) or transportation protocol (HTTP Web traffic).

Stateful Inspection Firewall

Filtering firewalls work on a packet at a time. They have no concept of "state" or "context" from one packet to next. A stateful inspection firewall is more sophisticated and it maintains state information to provide better filtering

Personal Firewall

A personal firewall is an application program that runs on a workstation or a PC to block unwanted traffic on a single workstation. A personal firewall can be configured to enforce a certain policy. For example, a user may decide that certain sites (for example a computer on a company network) is trustworthy and the firewall should allow traffic from only those sites. It is useful to combine a virus scanner with a personal firewall. For example, a firewall can direct all incoming email to a virus scanner, which examines every attachment the moment it reaches a particular host.

Application Proxy Gateway

An application proxy gateway is also called a *bastion host.* It is a firewall that simulates the proper effects of an application so that the application will receive only requests to act properly. The proxies on a firewall can be tailored to specific requirements such as logging details about the access. A proxy can demand strong authentication such as name, password and challenge-response.

Guard

A guard is another form of a sophisticated firewall. It receives protocol data units, interprets them and passes through the same or different protocol data units that achieve either the same result or a modified result. The guard decides what services to perform on user's behalf in accordance with its available knowledge. The following example illustrates the use of a guard. A university wants all students to restrict the size of email messages to a certain number of words or characters. Although, this rule can be implemented by modifying email handlers, it is more easily done by monitoring the common point through which all email flows.

6.3   Limitations of Firewalls

Firewalls do not offer complete solutions to all computer security problems. A firewall can only protect the perimeter of its environment against attacks from outsiders. The following are some of the important points about firewall based protection

Firewalls can only protect if they control the entire perimeter. Even if one inside host connects to an outside address by a modem, the entire inside net can be vulnerable through the modem and its host.

Firewalls are the most visible parts of a network and therefore they are the most attractive target for attacks..

Firewalls exercise only minor control over the content of the packets that are admitted inside the network. Therefore inaccurate data or malicious code must be controlled by other means inside the parameter.

7.   BASICS OF INTRUSION DETECTION SYSTEMS

Perimeter controls such as a firewall, or authentication and access control act as the first line of defense. However, prevention is not a complete security solution. Intrusion Detection systems complement these preventive controls by acting as the next line of defense. An IDS is a sensor, like a smoke detector that raises an alarm if specific things occur. Intrusion Detection is the process of identifying and responding to malicious activities targeted at computing and network resources. It involves technology, people and tools. An Intrusion Detection System basically monitors and collects data from a target system that should be protected, processes and correlates the gathered information and initiate responses when an intrusion is detected.

8.   CONCLUSIONS

Computer security ensures the confidentiality, integrity and availability of computing resources: hardware, software, network and data. These components have vulnerabilities and people exploit these vulnerabilities to stage attacks against these resources.

In this chapter we have discussed some of the salient features of security in networks and distributed applications. Since the world is becoming connected by computers the significance of network security will continue to grow. When a network and its components are designed and architectured well, the resulting system is quite resilient to attacks.

A lot of work is being done to enhance computer security. Products from vendor companies will lead to more secure boxes. There is a lot of research interests in the area of authentication, access control and authorizations. Another challenge for security is that networks are pervasive: cell phones, personal digital assistants and other consumer appliances are being connected. New applications lead to a new protocol development. There is a need to make sure that these protocols are tested for security flaws and that security measures are incorporated as needed. Intrusion Detection Systems and Firewalls have become popular products to secure networks. In the future, security of mobile code and web services will become an important issue as remote updates and patches become popular.

References

1. Pfleeger C.P. and Pfleeger S.L, "Security in Computing", Third Edition, Published by Prentice Hall, 2003
2. Bishop Matt, "Computer Security Art and Science", Addison-Wesley 2003, ISBN 0-201-44099-7
3. M. D. Abrams, S. Jajodia, and H. J. Podell, eds., Information Security: An Integrated Collection of Essays. IEEE Computer Society Press, 1995
4. A.J. Menezes, P.C. van Oorschot, and S.A. Vanstone, Handbook of Applied Cryptography, CRC Press, 1996
5. E. Amoroso, Fundamentals of Computer Security Technology, Prentice Hall, 1994
6. C.Kaufman, R.Perlman, and M.Speciner. Network Security: Private Communication in a Public World. 2nd ed. Prentice Hall, 2002
7. R.Anderson, Security Engineering, John Wiley and Sons 2001

8. Stallings W, "Network Security Essentials", Prentice Hall 2003, ISBN 0-13-035128-8
9. W. Cheswick, S.M. Bellovin, A. Rubin, "Firewalls and Internet Security", Addison Wesley, ISBN 0-201-63466-X, 2003

Chapter 3

INTRUSION DETECTION SYSTEMS

Anoop Singhal

Abstract: This chapter provides an overview of the state of the art in intrusion detection systems. Intrusion detection systems are software/hardware components that monitor systems and analyze the events for intrusions. This chapter first provides a taxonomy of intrusion detection systems. Second, architecture of IDS and their basic characteristics are presented. Third, a brief survey of different IDS products is discussed. Finally, significant gaps and direction for future work is discussed

Key words: intrusion, signatures, anomaly, data mining

Intrusion Detection is the process of identifying and responding to malicious activity targeted at computing and networking resources. It is a device, typically another computer that monitors activities to identify malicious or suspicious events. An IDS receives raw input from sensors, analyzes those inputs and then takes some action.

Since the cost of information processing and Internet accessibility is dropping, more and more organizations are becoming vulnerable to a wide variety of cyber threats. According to a recent survey by CERT, the rate of cyber attacks has been doubling every year in recent times. Therefore, it has become increasingly important to make our information systems, especially those used for critical functions such as military and commercial purpose, resistant to and tolerant of such attacks. Intrusion Detection Systems (IDS) are an integral part of any security package of a modern networked information system. An IDS detects intrusions by monitoring a network or system and analyzing an audit stream collected from the network or system to look for clues of malicious behavior.

1.   CLASSIFICATION OF INTRUSION DETECTION SYSTEMS

Intrusion Detection Systems can be described in terms of three functional components:

1. Information Sources: The different sources of data that are used to determine the occurrence of an intrusion. The common sources are network, host and application monitoring.
2. Analysis: This part deals with techniques that the system uses to detect an intrusion. The most common approaches are misuse detection and anomaly detection
3. Response: This implies the set of actions that the system takes after it has detected an intrusion. The set of actions can be grouped into active and passive actions. An active action involves an automated intervention whereas a passive action involves reporting IDS alerts to humans. The humans are in turn expected to take action.

Information Sources

Some IDSs analyze network packets captured from network bones or LAN segments to find attackers. Other IDSs analyze information generated by operating system or application software for signs of intrusion.

Network Based IDS

A network based IDS analyzes network packets that are captured on a network. This involves placing a set of traffic sensors within the network. The sensors typically perform local analysis and detection and report suspicious events to a central location. The majority of commercial intrusion detection systems are network based. One advantage of a network based IDS is that a few well placed network based IDS can monitor a large network. A disadvantage of a network based IDS is that it cannot analyze encrypted information. Also, most network based IDS cannot tell if an attack was successful, they can only detect that an attack was started.

Host Based IDS

A host based IDS analyzes host-bound audit sources such as operating system audit trails, system logs or application logs. Since host based systems directly monitor the host data files and operating system processes, they can determine exactly which host resources are targets of a particular attack. Due to the rapid development of computer networks, traditional single host intrusion detection systems have been modified to monitor a number of hosts on a network. They transfer the monitored information from multiple monitored hosts to a central site for processing. These are termed as distributed intrusion detection systems. One advantage of a host based IDS is that it can "observe" the outcome of an attempted attack, as it can directly access and monitor the data files and system processes that are usually targeted by attacks. A disadvantage of a host based IDS is that it is harder to manage and it is more vulnerable to attacks.

Application Based IDS

Application based IDS are a special subset of host based IDS that analyze the event that occur within a software application. The application log files are used to observe the events. One advantage of application based IDS is that they can directly monitor the interaction between a user and an application which allows them to trace individual users.

IDS Analysis

There are two primary approaches to analyze events to detect attacks: misuse detection and anomaly detection. Misuse detection is used by most commercial IDS systems and the analysis targets something that is known to be bad. Anomaly detection is one in which the analysis looks for abnormal forms of activity. It is a subject of great deal of research and is used by a limited number of IDS.

Misuse Detection

This method finds intrusions by monitoring network traffic in search of direct matches to known patterns of attack (called signatures or rules).

This kind of detection is also sometimes called "signature based detection". A common form of misuse detection that is used in commercial products specifies each pattern of events that corresponds to an attack as a separate signature. However, there are more sophisticated approaches called state based analysis that can leverage a single signature to detect a group of attacks.

A disadvantage of this approach is that it can only detect intrusions that match a pre-defined rule. The set of signatures need to be constantly updated to keep up with the new attacks. One advantage of these systems is that they have low false alarm rates.

Anomaly Detection

In this approach, the system defines the expected behavior of the network in advance. The profile of normal behavior is built using techniques that include statistical methods, association rules and neural networks. Any significant deviations from this expected behavior are reported as possible attacks. The measures and techniques used in anomaly detection include:

- Threshold Detection: In this kind of IDS certain attributes of user behavior are expressed in terms of counts, with some level established as permissible. Some examples of these attributes include number of files accessed by a user in a given period, the number of failed attempts to login to the system, the amount of CPU utilized by a process.
- Statistical Measures: In this case the distribution of profiled attributes is assumed to fit a pattern.
- Other Techniques: These include data mining, neural networks, genetic algorithms and immune system models.

In principle, the primary advantage of anomaly based detection is the ability to detect novel attacks for which signatures have not been defined yet. However, in practice, this is difficult to achieve because it is hard to obtain accurate and comprehensive profiles of normal behavior. This makes an anomaly detection system generate too many false alarms and it can be very time consuming and labor intensive to sift through this data.

Response Options for IDS

After an IDS has detected an attack, it generates responses. Commercial IDS support a wide range of response options, categorized as active responses, passive responses or a mixture of two.

Active Responses

Active responses are automated actions taken when certain types of intrusions are detected. There are three categories of active responses.

Collect additional information

The most common response to an attack is to collect additional information about a suspected attack. This might involve increasing the level of sensitivity of information sources for example turn up the number of events logged by an operating system audit trail or increase the sensitivity of a network monitor to capture all the packets. The additional information collected can help in resolving and diagnosing whether an attack is taking place or not.

Change the Environment

Another kind of active response is to halt an attack in progress and block subsequent access by the attacker. Typically, an IDS accomplishes this by blocking the IP address from which the attacker appears to be coming.

Take Action Against the Intruder

Some folks in the information warfare area believe that the first action in active response area is to take action against the intruder. The most aggressive form of this response is to launch an attack against the attacker's host or site.

Passive Responses

Passive IDS responses provide information to system users and they assume that human users will take subsequent action based on that information. Alarms and notifications are generated by an IDS to inform users when an attack is detected. The most common form of an alarm is an on screen alert or a popup window. Some commercial IDS systems are designed to generate alerts and report them to a network management system using SNMP traps. They are then displayed to the user via the network management consoles.

2.   INTRUSION DETECTION ARCHITECTURE

Figure 1 shows different components of IDS. They are briefly described below.

Target System: The System that is being analyzed for intrusion detection is considered as the target system. Some examples of target systems are corporate intranets and servers.

Feed: A feed is an abstract notion of information from the target system to the intrusion detection system. Some examples of a feed are system log files on a host machine or network traffic and connections.

Processing: Processing is the execution of algorithms designed to detect malicious activity on some target system. These algorithms can either use signature or some other heuristic techniques to detect the malicious activity. The physical architecture of the machine should have enough CPU power and memory to execute the different algorithms.

Knowledge Base: In an intrusion detection system, knowledge bases are used to store information about attacks as signatures, user and system behavior as profiles. These knowledge bases are defined with appropriate protection and capacity to support intrusion detection in real time.

Storage: The type of information that must be stored in an intrusion detection system will vary from short termed cached information about an ongoing session to longer term event related information for trend analysis. The storage capacity requirements will grow as networks start working at higher speeds.

Alarms/directives: The most common response of an intrusion detection system is to send alarms to a human analyst who will then analyze it and take proper action. However, the future trend is for IDS to take some action (e.g. update the access control list of a router) to prevent further damage. As this trend continues, we believe that intrusion detection will require messaging architectures for transmitting information between components.

Such messaging is a major element of the Common Intrusion Detection Framework.

GUI/operator interface: Proper Display of alarms from an IDS are usually done using a Graphical User Interface. Most commercial IDS have a fancy GUI with capabilities for data visualization and writing reports.

Communications infrastructure: Different components of an IDS and different IDS communicate using messages. This infrastructure also requires protection such as encryption and access control.

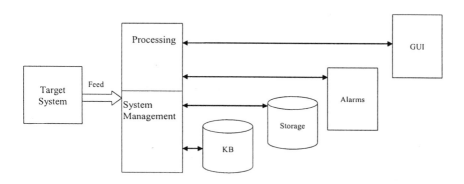

Figure 1: Intrusion Detection Architecture

3.   IDS PRODUCTS

This section presents some of the research and commercial IDS products.

3.1   Research Products

EMERALD

Event Monitoring Enabling Responses to Anomalous Live Disturbances is a research tool developed by SRI International. They have explored issues in intrusion detection associated with both deviations from normal user behavior (anomalies) and known intrusion patterns (signatures).

NetStat

This is a research tool produced by the University of California at Santa Barbara. It explores the use of state-transition analysis to detect real time intrusions.

Bro

Bro is a research tool developed by the Lawrence Livermore National Laboratory. The main design goals of Bro were
a) High Load Monitoring
b) Real Time Notification
c) Separating Mechanism from Policy
d) Ability to protect against attacks on the IDS

3.2   Commercial Products

This section gives examples of some of the commercial products.

NetProwler

This is a product from Axent Corporation. It supports both host based and network based detection. NetProwler provides signatures for a wide variety of operating system and application attacks. It allows a user to build customized signature profiles using a signature definition wizard. Examples of attack signatures that NetProwler supports include denial of service, unauthorized access, vulnerability probes and suspicious activity that is counter to company policies.

NetRanger

This is a product from Cisco Systems. It operates in real time and is scalable to enterprise level. A NetRanger system is composed of Sensors and one or more Directors that are connected by a communication system. In addition to providing many standard attack signatures, NetRanger provides the ability for the user to define their own customized signatures. In response to an attack, the Sensor can be configured with several options that include generating an alarm, logging the alarm event and denying further network access.

The Director provides a centralized management support for the NetRanger system. It allows the cabability to remotely install new signatures into the Sensors. The Director also provides a centralized collection and analysis of alert data. The status of Sensors can be monitored by the Director using a color coded scheme.

RealSecure

This is a product from ISS. It uses a three level architecture consisting of a network-based engine, a host based engine and an administrator's module. The network recognition engine runs on dedicated workstations to provide intrusion detection and response. Each network recognition engine watches the packet traffic traveling over a specific network segment for attack signatures. If it detects unauthorized activity, it can respond by terminating the connection, sending email or pager alerts, reconfiguring the firewalls or taking some user definable action.

The host based recognition engine is a host resident complement to the network recognition engine. It analyzes host logs to recognize attacks, determines whether the attack was successful or not and provides other forensic information that is not available in a real time environment.

All recognition engines report to and are configured by the administrative module, a management console that monitors the status of any number of UNIX and Windows NT recognition engines. This results in a comprehensive protection that is easily configured and administered from a single location.

3.3   Public Domain Tools

TripWire

This is a file integrity assessment tool that was originally developed at Purdue University. Tripwire is different from others as it detects changes in the file system of the monitored system. Tripwire comes in both commercial and free versions.

Tripwire computes checksums or cryptographic signatures of files. It can be configured to report all changes in the monitored file system. For example, it can check if system binaries have been modified, if syslog files have shrunk or if security settings have unexpectedly changed. It can be configured to perform integrity checks at regular intervals and to provide information to system administrators to implement recovery if tampering has occurred.

SNORT

SNORT is an open source NIDS that uses a combination of rules and preprocessors to analyze traffic. SNORT is easy to configure allowing users to create their own signatures and to alter the base functionality using plugins. Snort has evolved from a simple network management tool to a world-class enterprise distributed intrusion detection system. Snort detects suspicious traffic by using signature matching. Snort signatures are written and released by the Snort community within hours of the announcement of a new security exposure. It has the largest and most comprehensive collection of attack signatures for any IDS.

SNORT uses output plug-ins to store the output of its detection engine. It's outputting functionality is modular and provides different formats (e.g. XML, Relational Database. Text logfile and so on) to store the output. SNORT also provides a GUI to view the alerts. ACID is a Web application that reads intrusion data stored in a database and presents it in a browser in a human friendly format. ACID includes a charting component that is used to create statistics and graphs.

Network Flight Recorder

This is a network based IDS that was previously available in both a commercial version and a public domain version. NFR includes a complete programming language, called N, designed for packet analysis. Filters are written in this language which is compiled into byte code and interpreted by the execution engine. Programs can be written in N to perform pattern matching. Also, functions are provided to store the alert data into a database and perform back end analysis. Some examples of backend analysis are *histogram* and *list*. *Histogram* provides a facility for capturing data in a multi dimensional matrix. The system can be programmed to generate alerts based on the counts in different cells. The *list* functions allows records to be stored in a chronological order to store historical information.

NFR also provides *query* back ends that allow you to analyze the data. Query back ends have their own CGI interface and they also provide graphical functions for data visualization.

3.4 Government Off-the Shelf (GOTS) Products

CIDDS (Common Intrusion Detection Director System)

This was supported by the Air Force Information Warfare Center. CIDDS receives near real time connections data and associated events from Automated Security Incident Measurement (ASIM) Sensor host machines and selected other IDS tools. It stores this data on a local database and allows for detailed correlation and analysis by human analyst. Various uses of this data include

- Detecting potentially intrusive activities
- Detecting those activities that target specific machines
- Trend analysis for historical purposes

CIDDS provides the ASIM system with a centralized data storage and analysis capability.

ASIM Sensor

ASIM Sensor is a promiscuous data packet sniffer and analyzer. It consists of a suite of compiled C code and Java language programs. Real-time ASIM identifies strings and services that could indicate attempts at unauthorized access.

4. TYPES OF COMPUTER ATTACKS COMMONLY DETECTED BY IDS

Three type of computer attacks are commonly detected by IDS: system scanning, denial of service (DOS) and system penetration. These attacks can be launched locally, on the attacked machine or remotely using a network to access the target.

4.1 Scanning Attacks

A scanning attack occurs when an attacker probes a target network or system by sending different kinds of packets. From the responses received, the attacker can learn many of the system characteristics and vulnerabilities. Some of the information that the responses can provide are

- Topology of the target network
- Active hosts and operating systems on those hosts
- The different applications that are running on the host and their version numbers

Various tools that are used to perform these activities are: network mappers, port mappers, port scanners and vulnerability scanning tools.

4.2   Denial of Service Attacks

Denial of Service (DOS) attacks attempt to slow or shut down targeted network systems or services. There are two main types of DOS attacks: flaw exploitation and flooding.

4.2.1   Flaw Exploitation DOS Attacks

In this kind of an attack the attacker exploits a flaw in the target system's software. An example of such a processing failure is the "ping of death" attack. This attack involves sending a large ping packets that the target system cannot handle and it will result in a crash.

4.2.2   Flooding DOS Attack

A flooding attack is one in which the attacker sends more information than what the target can handle which results in an exhaustion of system resources. One example of this attack is the "SYN Flood" attack. The term distributed DOS attack is used where the attacker uses multiple computers to launch an attack.

4.3   Penetration Attacks

These attacks involve unauthorized acquisition or alteration of a system resource. Consider these as integrity and control violations. Some examples of penetration attacks are

User to Root: A local user on a host gains root access

Remote to User: An attacker on a network gains access to a user account on the target host

Remote to Root: An attacker on the network gains complete control of the target host

5.   SIGNIFICANT GAPS AND FUTURE DIRECTIONS FOR IDS

This section discusses significant gaps in the current IDS products.

a) Historical Data Analysis: As networks are getting large and complex, security officers that are responsible for managing these networks need tools that help in historical data analysis, generating reports and doing trend analysis on alerts that were generated in the past. Current IDS often generate too many *false alerts* due to their simplistic analysis. The storage management of alerts from IDS for a complex network is a challenging task.

b) Support for Real Time Alert Correlation: Intrusion correlation refers to interpretation, combination and analysis of information from several sensors. For large networks, sensors will be distributed and they will send their alerts to one central place for correlation processing. There is a need for this information to be stored and organized efficiently at the correlation center. Also, traditional IDSs focus on low level alerts and they do not group them even if there is a logical connection among them. As a result, it becomes difficult for human users to understand these alerts and take appropriate actions. It has been reported that for a typical network "users are encountering 10 to 20,000 alerts per sensor per day". Therefore, there is a need to store these alerts efficiently and group them to construct attack scenarios.

c) Heterogeneous Data Support: In a typical network environment, there are multiple audit streams from diverse cyber sensors 1) raw network traffic data 2) netflow data 3) system calls 4) output alerts from an IDS and so on. It is important to have an architecture that can integrate these data sources into a unified framework. together so that an analyst can access it in real time. Since current IDS are not perfect they produce a lot of *false alarms*. There is a need for efficient querying techniques for a user to verify if an alert is genuine by correlating it with the input audit data.

d) Forensic Analysis: With the rapidly growing theft and unauthorized destruction of computer based information, the frequency of prosecution is rising. To support prosecution, electronic data must be captured and stored in such a way that it provides legally acceptable evidence.

e) Feature extraction from Network Traffic Data and Audit Trails: For each type of data that needs to be examined (network packets, host event logs, process traces etc.) data preparation and feature extraction is currently a challenging task. Due to large amounts of data that needs to be prepared for the operation of IDS system, this becomes expensive and time consuming.

f) Data Visualization: During attack, there is a need for the system administrator to graphically visualize the alerts and respond to them. There is also a need to filter and view alerts, sorted according to priority, sub-net or time dimensions.

In the next chapter we will describe how data warehousing and data mining techniques can solve some of these problems in IDS Applications.

6.   CONCLUSIONS

The market for IDS and vulnerability assessment products has grown drastically in the last few years. While IDS research is maturing, commercial IDS products have become stable. Some commercial IDS systems have been blamed for large number of false alarm rates, awkward user interfaces and difficult to use. However, the strong commercial demand for IDS has forced the commercial IDS vendors to solve these problems in a timely manner. Furthermore, it is likely that certain IDS capabilities will become core features of network infrastructure such as routers, bridges and switches.

References

1.   There are several books on IDS including:
- Bace, Rebecca G., Intrusion Detection, Macmillan Technical Publishing, 2000
- Amoroso, Edward G., Intrusion Detection: An Introduction to Internet Surveillance, Intrusion.net 1999.
- Northcutt, Stephen, Network Intrusion Detection: An Analyst's Handbook, New Riders, 1999
2.   For an overview of IDS and their capabilities read the white paper "An Introduction to Intrusion Detection Assessment", at http://www.icsa.net/services/consortia/intrusion/intrusion.pdf
3.   Snort is a lightweight network intrusion detection system, which can perform a variety of logging and analysis functions on IP networks. The URL for both Snort and attack signatures is http://www.snort.otg
4.   RealSecure IDS, http://www.iss.net
5.   CERIAS at Purdue University has produced several network security tools. The URL is http://www.cerias.purdue.edu

Chapter 4

DATA MINING FOR INTRUSION DETECTION

Anoop Singhal

Abstract: Data Mining Techniques have been successfully applied in many different fields including marketing, manufacturing, fraud detection and network management. Over the past years there is a lot of interest in security technologies such as intrusion detection, cryptography, authentication and firewalls. This paper discusses the application of Data Mining techniques to computer security. Conclusions are drawn and directions for future research are suggested.

Key words: anomaly detection, correlation, association rules, classification

1. INTRODUCTION

Intrusion detection is the process of monitoring and analyzing the events occurring in a computer system in order to detect signs of security problems. Over the past several years, intrusion detection and other security technologies such as cryptography, authentication and firewalls have increasingly gained in importance. There is a lot of interest in applying data mining techniques to intrusion detection. This chapter gives a critical summary of data mining research for intrusion detection. We provide a survey of research projects that apply data mining techniques to intrusion detection. We then suggest new directions for research and then give our conclusions.

2.   DATA MINING FOR INTRUSION DETECTION

Recently, there is a great interest in application of data mining techniques to intrusion detection systems. The problem of intrusion detection can be reduced to a data mining task of classifying data. Briefly, one is given a set of data points belonging to different classes (normal activity, different attacks) and aims to separate them as accurately as possible by means of a model. This section gives a summary of the current research project in this area.

2.1   Adam

The ADAM project at George Mason University [1], [2] is a network-based anomaly detection system. ADAM learns normal network behavior from attack-free training data and represents it as a set of association rules, the so called profile. At run time, the connection records of past delta seconds are continuously mined for new association rules that are not contained in the profile.

ADAM is an anomaly detection system. It is composed of three modules: a preprocessing engine, a mining engine and a classification engine. The preprocessing engine sniffs TCP/IP traffic data and extracts information from the header of each connection according to a predefined schema. The mining engine applies mining association rules to the connection records. It works in two modes: training mode and detecting mode. In training mode, the mining engine builds a profile of the users and systems normal behavior and generates association rules that are used to train the classification engine. In detecting mode, the mining engine mines unexpected association rules that are different from the profile. The classification engine will classify the unexpected association rules into normal and abnormal events. Some abnormal events can be further classified as attacks. Although mining of association rules has used previously to detect intrusions in audit trail data, the ADAM system is unique in the following ways:

It is on-line; it uses an incremental mining (on-line mining) which does not look at a batch of TCP connections, but rather uses a sliding window of time to find the suspicious rules within that window.

It is an anomaly detection system that aims to categorize using data mining the rules that govern misuse of a system. For this, the technique builds, apriori, a profile of "normal" rules, obtained by mining past periods of time in which there were no attacks. Any rule discovered during the on-line mining that also belongs to this profile is ignored, assuming that it corresponds to a normal behavior.

Figures 1 and 2 show the basic architecture of ADAM. ADAM performs its task in two phases. In the training phase, ADAM uses a data stream for which it knows where the attacks are located. The attack free parts of the stream are fed into a module that performs off-line association rules discovery. The output of this module is a profile of rules that we call "normal" i.e. it provides the behavior during periods when there are no attacks. The profile along with the training data set is also fed into a module that uses a combination of dynamic, on line algorithm for association rules, whose output consists of frequent item sets that characterize attacks to the system. These item sets are used as a classifier or decision tree. This whole phase takes place off-line before we use the system to detect attacks.

The second phase of ADAM in which we actually detect attacks is shown in the figure below. Again, the on-line association rules mining algorithm is used to process a window of current connections. Suspicious connections are flagged and sent along with their feature vectors to the trained classifier, where they are labeled as attacks, false alarms or unknown. When, the classifier labels connections as false alarms, it is filtering them out of the attacks set and avoiding passing these alerts to the security officer. The last class, i.e. unknown is reserved for the events whose exact nature cannot be confirmed by the classifier. These events are also considered as attacks and they are included in the set of alerts that are passed to the security officer.

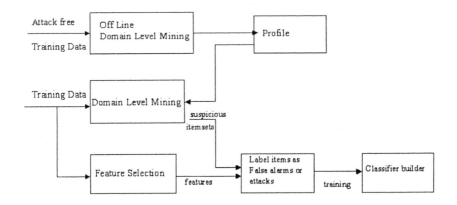

Figure 1: The Training Phase of ADAM

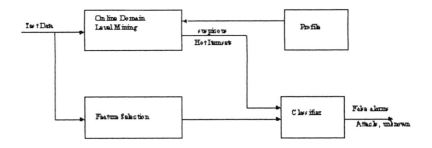

Figure 2: The Intrusion Detection Phase of ADAM

2.2   Madam ID

The MADAM ID project at Columbia University [7], [8] has shown how data mining techniques can be used to construct an IDS in a more systematic and automated manner. Specifically, the approach used by MADAM ID is to learn classifiers that distinguish between intrusions and normal activities. Unfortunately, classifiers can perform really poorly when they have to rely on attributes that are not predictive of the target concept. Therefore, MADAM ID proposes association rules and frequent episode rules as means to construct additional more predictive attributes. These attributes are termed as *features*.

We will describe briefly how MADAM ID is used to construct network based misuse detection systems. First all network traffic is preprocessed to create *connection records*. The attributes of connection records are intrinsic connection characteristics such as source host, the destination host, the source and destination posts, the start time, the duration, header flags and so on. In the case of TCP/IP networks, connection records summarize TCP sessions.

The most important characteristic of MADAM ID is that it *learns* a misuse detection model from examples. In order to use MADAM ID, one needs a large set of connection records that have already been classified into "normal records" or some kind of attacks. MADAM ID proceeds in two steps. In the first step it does *feature construction* in which some additional features are constructed that are considered useful for doing the analysis. One example for this step is to calculate the count of the number of connections that have been initiated during the last two seconds to the same destination host as the current host. The feature construction step is followed by the *classifier learning* step. It consists of the following process:

1. The training connection records are partitioned into two sets, namely *normal connection records* and *intrusion connection records*.
2. Association rules and frequent episode rules are mined separately from the normal connection records and from the intrusion connection records. The resulting patterns are compared and all patterns that are exclusively contained in the intrusion connection records are collected to form *the intrusion only* patterns.
3. The intrusion only patterns are used to derive additional attributes such as count or percentage of connection records that share some attribute values with the current connection records.

4. A classifier is learned that distinguishes normal connection records from intrusion connection records, This classifier is the end product of MADAM ID.

2.3   Minds

The MINDS project [4] [6] at University of Minnesota uses a suite of data mining techniques to automatically detect attacks against computer networks and systems. Their system uses an anomaly detection technique to assign a score to each connection to determine how anomalous the connection is compared to normal network traffic. Their experiments have shown that anomaly detection algorithms can be successful in detecting numerous novel intrusions that could not be identified using widely popular tools such as SNORT.

Input to MINDS is Netflow data that is collected using Netflow tools. The netflow data contains packet header information i.e. they do not capture message contents. Netflow data for each 10 minute window which typically results in 1 to 2 million records is stored in a flat file. The analyst uses MINDS to analyze these 10 minute data files in a batch mode. The first step in MINDS involves constructing features that are used in the data mining analysis. Basic features include source IP address and port, destination IP address and port, protocol, flags, number of bytes and number of packets. Derived features include time-window and connection window based features. After the feature construction step, the data is fed into the MINDS anomaly detection module that uses an outlier detection algorithm to assign an anomaly score to each network connection. A human analyst then has to look at only the most anomalous connections to determine if they are actual attacks or other interesting behavior.

MINDS uses a density based outlier detection scheme for anomaly detection. The reader is referred to [4] for a more detailed overview of their research. MINDS assigns a degree of being an outlier to each data point which is called the local outlier factor (LOF). The output of the anomaly detector contains the original Netflow data with the addition of the anomaly score and relative contribution of the different attributes to that score. The analyst typically looks at only the top few connections that have the highest anomaly scores. The researchers of MINDS have their system to analyze the University of Minnesota network traffic. They have been successful in detecting scanning activities, worms and non standard behavior such as policy violations and insider attacks.

2.4   Clustering of Unlabeled ID

Traditional anomaly detection systems require "clean" training data in order to learn the model of normal behavior. A major drawback of these systems is that clean training data is not easily available. To overcome this weakness, recent research has investigated the possibility of training anomaly detection systems over noisy data [11]. Anomaly detection over noisy data makes two key assumptions about the training data. First, the number of normal elements in the training data is assumed to be significantly larger than the number of anomalous elements. Secondly, anomalous elements are assumed to be qualitatively different from normal ones. Then, given that anomalies are both rare and different, they are expected to appear as outliers that stand out from the normal baseline data. Portnoy et al. [11] apply clustering to the training data. Here the hope is that intrusive elements will bundle with other intrusive elements whereas normal elements will bundle with other normal ones. Moreover, as intrusive elements are assumed to be rare, they should end up in small clusters. Thus, all small clusters are assumed to contain intrusions/anomalies, whereas large clusters are assumed to represent normal activities. At run time, new elements are compared against all clusters and the most similar cluster determines the new element's classification as either "normal" or "intrusive".

2.5   Alert Correlation

Correlation techniques from multiple sensors for large networks is described in [9], [10]. A language for modeling alert correlation is described in [3]. Traditional IDS systems focus on low level alerts and they raise alerts independently though there may be a logical connection between them. In case of attacks, the number of alerts that are generated become unmanageable. As a result, it is difficult for human users to understand the alerts and take appropriate actions. Ning et al. present a practical method for constructing attack scenarios through alert correlation, using prerequisites and consequences of intrusions. Their approach is based on the observation that in a series of attacks, alerts are not isolated, but related as different stages, with earlier stages preparing for the later ones. They proposed a formal framework to represent alerts with their prerequisites and consequences using the concept of *hyper-alerts*. They evaluated their approach using the 2000 DARPA intrusion detection scenario specific datasets.

3.   CONCLUSIONS AND FUTURE RESEARCH DIRECTIONS

In this chapter, we reviewed the application of data mining techniques to the area of computer security. Data mining is primarily being used to detect intrusions rather than to discover new knowledge about the nature of attacks. Moreover, most research is based on strong assumptions that complicate building of practical applications. First, it is assumed that labeled training data is readily available, and second it is assumed that this data is of high quality. Different authors have remarked that in many cases, it is not easy to obtain labeled data. Even if one could obtain labeled training data by simulating intrusions, there are many problems with this approach. Additionally, attack simulation limits the approach to the set of known attacks. We think that the difficulties associated with the generation of high quality training data will make it difficult to apply data mining techniques that depend on availability of high quality labeled training data. Finally, data mining in intrusion detection focuses on a small subset of possible applications. Interesting future applications of data mining might include the discovery of new attacks, the development of better IDS signatures and the construction of alarm correlation systems.

For future research, it should be possible to focus more on the KDD process and detection of novel attacks. It is known that attackers use a similar strategy to attack in the future as what they used in the past. The current IDSs can only detect a fraction of these attacks. There are new attacks that are hidden in the audit logs, and it would be useful to see how data mining can be used to detect these attacks.

Data mining can also be applied to improve IDS signatures. IDS vendors can run their systems in operational environment where all alarms and audit logs are collected. Then, data mining can be used to search for audit log patterns that are closely related with particular alarms. This might lead to new knowledge as to why false positives arise and how they can be avoided.

Finally, data mining projects should focus on the construction of alarm correlation systems. Traditional intrusion detection systems focus on low level alerts and they raise alerts independently even though there is a logical connection among them. More work needs to be done on alert correlation techniques that can construct "attack strategies" and facilitate intrusion analysis. One way is to store data from multiple sources in a data warehouse and then perform data analysis. Alert correlation techniques will have several advantages. First, it will provide a high level representation of the alerts along with a temporal relationship of the sequence in which these alerts occurred. Second, it will provide a way to distinguish a true alert from a false alert. We think that true alerts are likely to be correlated with other alerts whereas false alerts will tend to be random and, therefore, less likely to be related to other alerts. Third, it can be used to anticipate the future steps of an attack and, thereby, come up with a strategy to reduce the damage.

References

1. Barbara D., Wu N., and Jajodia S., Detecting novel network intrusions using bayes estimators. In Proc. First SIAM Conference on Data Mining, Chicago, IL, April 2001.
2. Barbara D., Couto J., Jajodia S., and Wu N., Adam: Detecting Intrusions by Data Mining, In Proc. 2nd Annual IEEE Information Assurance Workshop, West Point, NY, June 2001.
3. Cuppens F. and Miege A., Alert Correlation in a Cooperative Intrusion Detection Framework, Proc. IEEE Symposium on Security and Privacy, May 2002.
4. Ertoz L., Eilertson E., Lazarevic A., Tan P., Dokes P., Kumar V., Srivastava J., Detection of Novel Attacks using Data Mining, Proc. IEEE Workshop on Data Mining and Computer Security, November 2003.
5. Han J and Kamber M., Data Mining: Concepts and Techniques, Morgan Kaufmann, August 2000
6. Kumar V., Lazarevic A., Ertoz L., Ozgur A., Srivastava J., A Comparative Study of Anomaly Detection Schemes in Network Intrusion Detection, Proc.Third SIAM International Conference on Data Mining, San Francisco, May 2003.
7. Lee W., Stolfo, S. J., and Kwok K. W. Mining audit data to build intrusion detection models. In Proc. Fourth International Conference on Knowledge Discovery and Data Mining, NewYork, 1998.
8. Lee W. and Stolfo S. J. Data Mining approaches for intrusion detection, In Proc. Seventh USENIX Security Symposium, San Antonio, TX, 1998.
9. Ning P., Cui Y., Reeves D. S., Constructing Attack Scenarios through Correlation of Intrusion Alerts, Proc. ACM Computer and Communications Security Conf., 2002.
10. Ning P., Xu D., earning Attack Strategies from Intrusion Alerts, Proc. ACM Computer and Communications Security Conf., 2003.
11. Portnoy L., Eskin E., Stolfo S. J., Intrusion Detection with unlabeled data using clustering. In Proceedings of ACM Workshop on Data Mining Applied to Security, 2001.

Chapter 5

DATA MODELING AND DATA WAREHOUSING TECHNIQUES TO IMPROVE INTRUSION DETECTION

Anoop Singhal

Abstract: This chapter describes data mining and data warehousing techniques that can improve the performance and usability of Intrusion Detection Systems (IDS). Current IDS do not provide support for historical data analysis and data summarization. This chapter presents techniques to model network traffic and alerts using a multi-dimensional data model and *star schemas*. This data model was used to perform network security analysis and detect denial of service attacks. Our data model can also be used to handle heterogeneous data sources (e.g. firewall logs, system calls, net-flow data) and enable up to two orders of magnitude faster query response times for analysts as compared to the current state of the art. We have used our techniques to implement a prototype system that is being successfully used at Army Research Labs. Our system has helped the security analyst in detecting intrusions and in historical data analysis for generating reports on trend analysis.

Key words: data warehouse, OLAP, data mining and analysis, star schema

1.   INTRODUCTION

This section describes the author's experience in designing a data warehousing system for historical data analysis and data summarization for analysts at the Center for Intrusion Detection in Army Research Labs.

Since the cost of information processing and Internet accessibility is dropping, more and more organizations are becoming vulnerable to a wide variety of cyber threats. According to a recent survey by CERT, the rate of cyber attacks has been doubling every year in recent times. Therefore, it has become increasingly important to make our information systems, especially those used for critical functions such as military and commercial purpose, resistant to and tolerant of such attacks. Intrusion Detection Systems (IDS) are an integral part of any security package of a modern networked information system. An IDS detects intrusions by monitoring a network or system and analyzing an audit stream collected from the network or system to look for clues of malicious behavior.

Intrusion Detection Systems generate a lot of alerts. There is a need to develop methods and tools that can be used by the system security analyst to understand the massive amount of data that is being collected by IDS, analyze and summarize the data and determine the importance of an alert. The problem is further complicated due the temporal variations. For instance, the "normal" number and duration of FTP connections may vary from morning to afternoon to evening. It may also vary depending on the class of users being considered.

In this chapter, we present data modeling, data visualization and data warehousing techniques that can drastically improve the performance and usability of Intrusion Detection Systems. Data warehousing and On Line Analytical Processing (OLAP) techniques can help the security officer in detecting attacks, monitoring current activities on the network, historical data analysis about critical attacks in the past, and generating reports on trend analysis. We present techniques for feature extraction from network traffic data and how a multi-dimensional data model or STAR schemas can be used to represent network traffic data and relate it to the corresponding IDS alerts.

This chapter is organized as follows. We first give a survey of research projects that apply data mining techniques to intrusion detection. Then we discuss the shortcomings in current systems. Section 3 presents a data architecture for IDS. Section 4 presents the data model followed by System Implementation. Finally section 5 gives the conclusions.

2.   BACKGROUND

Recently, there is a great interest in application of data mining techniques to intrusion detection systems [1]. The problem of intrusion detection can be reduced to a data mining task of classifying data. Briefly, one is given a set of data points belonging to different classes (normal activity, different attacks) and aims to separate them as accurately as possible by means of a model. This section gives a summary of the current research in this area.

1. MADAM ID: The MADAM ID project at Columbia University [2], [3] has shown how data mining techniques can be used to construct a IDS in a more systematic and automated manner.

2. ADAM: The ADAM project [4], [5] is a network-based anomaly detection system. ADAM learns normal network behavior from attack-free training data and represents it as a set of association rules, the so called profile. At run time, the connection records of past delta seconds are continuously mined for new association rules that are not contained in the profile.

3. MINDS: The MINDS project [6], [7] at University of Minnesota uses a suite of data mining techniques to automatically detect attacks against computer networks and systems. Their system uses an anomaly detection technique to assign a score to each connection to determine how anomalous the connection is compared to normal network traffic. Their experiments have shown that anomaly detection algorithms can be successful in detecting numerous novel intrusions that could not be identified using widely popular tools such as SNORT [].

4. Clustering of Unlabeled ID: Traditional anomaly detection systems require "clean" training data in order to learn the model of normal behavior. A major drawback of these systems is that clean training data is not easily available. To overcome these weakness, recent research has investigated the possibility of training anomaly detection systems over noisy data [8]

5. IDDM: The IDDM system [9] uses anomaly detection techniques for intrusion detection.

6. eBayes: The eBayes [10] system also uses anomaly detection for intrusion detection.

7. Alert Correlation: [11], [12] use correlation techniques to construct "attack scenarios" from low level alerts. [13] also describe a language for modeling alert correlation. [22], [23] describe probabilistic alert correlation. [24] describes use of attack graphs to correlate intrusion events.

3.   RESEARCH GAPS

Historical Data Analysis: As networks are getting large and complex, security officers that are responsible for managing these networks need tools that help in historical data analysis, generating reports and doing trend analysis on alerts that were generated in the past.   Current IDS often generate too many *false alerts* due to their simplistic analysis. The storage management of alerts from IDS for a complex network is a challenging task.

Support for Real Time Alert Correlation: Intrusion correlation refers to interpretation, combination and analysis of information from several sensors. For large networks, sensors will be distributed and they will send their alerts to one central place for correlation processing. There is a need for this information to be stored and organized efficiently at the correlation center. Also, traditional IDSs focus on low level alerts and they do not group them even if there is a logical connection among them. As a result, it becomes difficult for human users to understand these alerts and take appropriate actions. It has been reported that for a typical network "users are encountering 10 to 20,000 alerts per sensor per day". Therefore, there is a need to store these alerts efficiently and group them to construct attack scenarios [11], [12].

Heterogeneous Data Support: In a typical network environment, there are multiple audit streams from diverse cyber sensors 1) raw network traffic data 2) netflow data 3) system calls 4) output alerts from an IDS and so on. It is important to have an architecture that can integrate these data sources into a unified framework. together so that an analyst can access it in real time. Since current IDS are not perfect they produce a lot of *false alarms.* There is a need for efficient querying techniques for a user to verify if an alert is genuine by correlating it with the input audit data.

Forensic Analysis: With the rapidly growing theft and unauthorized destruction of computer based information, the frequency of prosecution is rising. To support prosecution, electronic data must be captured and stored in such a way that it provides legally acceptable evidence.

Feature extraction from Network Traffic Data and Audit Trails: For each type of data that needs to be examined (network packets, host event logs, process traces etc.) data preparation and feature extraction is currently a challenging task. Due to large amounts of data that needs to be prepared for the operation of IDS system, this becomes expensive and time consuming.

Data Visualization: During attack, there is a need for the system administrator to graphically visualize the alerts and respond to them. There is also a need to filter and view alerts, sorted according to priority, sub-net or time dimensions.

4.   A DATA ARCHITECTURE FOR IDS

In this section we describe a set of techniques that will considerably improve the performance of intrusion detection systems. The improvement is focused in the area of multi-dimensional data model that can be used to represent alerts and to detect new kinds of attacks. Techniques for feature extraction from network traffic data and alert correlation are also presented.

4.1   A Software Architecture and Data Model for Intrusion Detection

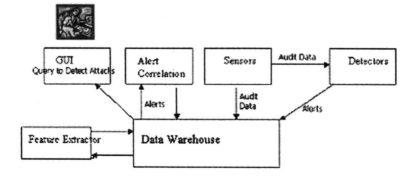

Figure 1: Data Architecture for Intrusion Detection System

Figure 1 shows an architecture diagram of our system. In a typical network environment there are many different audit streams that are useful for detecting intrusions. For example, such data includes network packets (headers, payload features), system logs on the host and system calls of processes on these machines. These types of data have different properties. Also, the detection models can vary. The most widely used detection model is a signature based system while data mining based approaches are also being explored. It is important to have an architecture that can handle any kind of data and different detection models. Our architecture supports the following components:

1. Real time components that includes sensors and detectors
2. A data warehouse component to store the data efficiently
3. Feature extraction component that reads the audit data from the data warehouse, extracts some features and computes some aggregates and then stores the information back in the data warehouse. These features are useful to the analysts to detect attacks.
4. Visualization engine that presents information to the analyst.

The proposed architecture has several advantages:

1. Modularity: All the data is stored in one central place and can be easily queried by the security analyst or the intrusion detection applications.
2. Support for multiple detectors: We have separated the sensor component from the detector component. This allows us to use a signature based detection engine and a data mining based detection engine on the same set of audit data.
3. Correlation of audit data from multiple sensors: Since the data from multiple sensors is stored in one central place, a detection engine can easily access the data from multiple sensors by executing a database query.
4. Reusability: Since the features extracted from the audit data are stored in one central place, they can be re-used by multiple applications to detect attacks.

Some more benefits for this software architecture and data warehouse for Network Fault Management and Provisioning Applications are discussed in [14] [15] [16].

4.2   Data Modeling for Historical Data Analysis Using STAR Schema:

In order to help the security officer or an analyst to decide whether an alert needs further investigation we plan to support the capability of querying and browsing a historical database. We model the alert data as a multidimensional dataset and borrow the model used in On Line Analytical Processing (OLAP). A popular abstraction for multidimensional data that is widely used in OLAP is the *data cube*. A cube is simply a multidimensional structure that contains at each point an aggregate value, i.e. the result of applying an aggregate function to an underlying relation.

In our case, the underlying relation is the alerts that are generated from an IDS. The alerts can be viewed as a multidimensional data. This schema is known as the *star schema*. In it, the main table is called the *fact table*. The attributes are the dimensions of the data. Examples of dimensions are *Time&Date, Duration, Sdinfo, Service, Attack.*   *Time&Date* contains information of date and time when the attack was staged. *Duration* records duration of the attack. *Sdinfo* describes the Source/Destination IP addresses and destination port information. This dimension encompasses a hierarchy which shows how this information can be aggregated to produce different views. Both, the source and destination IP addresses are composed of 4 bytes *Sip1Sip2Sip3Sip4* and *Dip1Dip2Dip3Dip4*. Dropping one or more of these fields produces a higher level view of the address. For example, *Sip1Sip2* corresponds to a series of domain of IP addresses each characterized by the first 2 bytes of the address. The *Service* dimension table contains the service name that was attacked and the class of service (e.g. TCP, UDP). The hierarchy for these dimensions are also shown. Similarly, the dimension table contains *Attack* contains both the name of the attack and its type (e.g. DOS, Probe). The dimension *Time&Date* presents different views of timing information. Finally, the dimension *Duration* contains the length of the attack. This can also be viewed as *long, medium* or *short.*

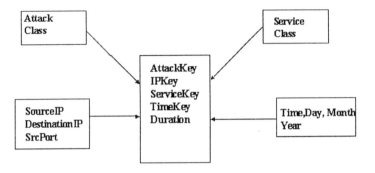

Figure 2: A Star Schema for the IDS Data Warehouse

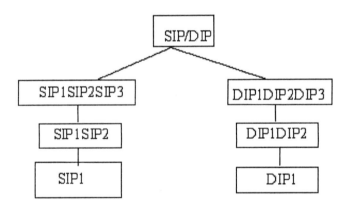

Figure 3: Dimension Hierarchy for IP Address

Using this schema, a corresponding cube would be a five dimensional structure in which cell contains aggregates of the operations measures. For instance, a cell could correspond to short duration attacks over the ftp service in the period 1 pm to 2 pm during Oct 20th 1998. Data cubes can be constructed by using SQL aggregation functions (COUNT, SUM, MIN, MAX). Cubes can be organized in a hierarchical manner. At the base of the hierarchy are the aggregates computed from the fact table. We call this *base data*. As data is consolidated into higher levels it is called *consolidated data*. For example, in our data cube, the base data could be cells that contain aggregates of measures per user, operation, time period and date. Higher levels of hierarchy can be specified in terms of classes of users (users in division W), coarser time periods (e.g. morning) and date consolidation (e.g. Sept. 2000).

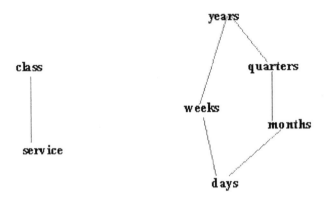

Figure 4: Dimension Hierarchy for Time and Services

4.3   Support for High Speed Drill Down Queries and Detection of Attacks/Virus/Worms

When an alert is generated by an IDS the analyst is interested to "drill down" and check the corresponding "raw network traffic" data to verify the alert. If the network traffic data is large (typically a Terabyte for 1 week of network traffic data) this can be time consuming. We describe techniques to organize the raw network traffic data using STAR schemas so that it is efficient to query it and link the raw network traffic data for the corresponding output alert. We use "bit map indexing" and "join indexing" techniques to speed up query processing. We have designed queries for security analysis of network traffic data. The following are some examples of security analysis:

a) Scanning Activity: Process one hour of data (17:00 –18:00 GMT on October 15th 2004) and look for all flows where the SYN flag was set and ACK/FIN flags are not set.

b) Recently, Sasser worm was released that scans port 445. To detect this worm a query was written to search for flows that scan for port 445. If the analyst is interested in internal machines that have been infected he can narrow the search to only those machines with destination port 445. A query was written that would generate the top ten source-destination IP pairs on destination port 445 for netflow data during a certain period of time.

c) Another security concern is denial of service attacks. One of the common network based denial of service attacks is SYN flooding. A query was written which was similar to those for worm detection to detect if a SYN flood has occurred. In this case we detected all source-destination IP pairs that have seen an excessive number of SYN packets.

d) Worm Detection: Recently, the MyDoom worm spread via an email attachment that created a backdoor on ports 3127-3198. After the release of this worm, scanning for this backdoor increased significantly. SQL queries were written to generate reports about the number of flows caused by this scanning in 10 minute intervals. The report shows that there is a sudden jump in the number of bytes transferred, even though the number of flows stayed constant.

4.4   Feature Extraction From Network Traffic Data

A number of data mining based IDS applications need to pre-process the network traffic data before they can do their analysis. For example, the preprocessing module of ADAM [4] generates a record for each connection from the header information of its packets based on the following schema:

R(TS, Src.IP, Src.Port, Dest.IP, Dest.Port, FLAG)

In this schema, TS represents the beginning time of a connection, Src.IP and Src.Port refer to source IP and port number respectively, while Dest.IP and Dest.Port represent the destination IP and port number. The attribute FLAG describes the status of a connection. This relation R is used for association mining. We store the connection records in the data warehouse so that they are available in one central place by several applications to do the analysis. Besides the basic features, we also store some derived features based on the window of time (number of bytes, number of packets, number of connections) that can be useful to detect attacks. These features are used to capture connections with similar characteristics (src-ip, dest-ip, src-port, dest-port, protocol) in the last T seconds, since typically DOS and scanning attacks involve hundreds of connections. A similar approach was used for constructing features in the KDDCup '99 data [20].

4.5   Help the Security Officer for Forensic Analysis

One of the important kind of analysis is forensic analysis. Currently forensic analysis of data is done manually. Computer experts have to search through large amounts of data, sometimes millions of records, individually and look for suspicious behavior. This is an extremely inefficient and expensive process. Since we can store all the historical data (net-flow data, system calls, fire-wall logs) in a data warehouse we can help the security officer in accessing all the records which are suspicious and possibly have some intrusions. The suspicious activity can then be labeled as either anomalous or normal using SQL statements to mark the appropriate data. Since all the data is stored in a data warehouse we can update the record and store it back in the database. Our database platform can be used to design Digital Forensics tools tailored to Information Warfare that can provide real time performance.

5. CONCLUSIONS

This chapter described data modeling and data warehousing techniques that drastically improve the performance and usability of Intrusion Detection Systems (IDS). Current IDS do not provide support for historical data analysis and data summarization. This chapter presented techniques to model network traffic and alerts using a multi-dimensional data model and *star schemas*. This data model was used to perform network security analysis and detect denial of service attacks. Our data model can also be used to handle heterogeneous data sources (e.g. firewall logs, system calls, net-flow data) and enable up to two orders of magnitude faster query response times for analysts as compared to the current state of the art. We have used our techniques to implement a prototype system that is being successfully used at Army Research Labs. Our system has helped the security analyst in detecting intrusions and in historical data analysis for generating reports on trend analysis

References

1. Singhal A. and Jajodia S., "Data Mining for Intrusion Detection", Published as a chapter in Data Mining Handbook, Kluwer, December 2004.
2. Lee W., Stolfo, S. J., and Kwok K. W. Mining audit data to build intrusion detection models. In Proc. Fourth International Conference on Knowledge Discovery and Data Mining, NewYork, 1998.
3. Lee W. and Stolfo S. J. Data Mining approaches for intrusion detection, In Proc. Seventh USENIX Security Symposium, San Antonio, TX, 1998.
4. Barbara D., Wu N., and Jajodia S., Detecting novel network intrusions using bayes estimators. In Proc. First SIAM Conference on Data Mining, Chicago, IL, April 2001.
5. Barbara D., Couto J., Jajodia S., and Wu N., Adam: Detecting Intrusions by Data Mining, In Proc. 2nd Annual IEEE Information Assurance Workshop, West Point, NY, June 2001.
6. Ertoz L., Eilertson E., Lazarevic A., Tan P., Dokes P., Kumar V., Srivastava J., Detection of Novel Attacks using Data Mining, Proc. IEEE Workshop on Data Mining and Computer Security, November 2003.

7. Kumar V., Lazarevic A., Ertoz L., Ozgur A., Srivastava J., A Comparative Study of Anomaly Detection Schemes in Network Intrusion Detection, Proc.Third SIAM International Conference on Data Mining, San Francisco, May 2003.

8. Portnoy L., Eskin E., Stolfo S. J., Intrusion Detection with unlabeled data using clustering. In Proceedings of ACM Workshop on Data Mining Applied to Security, 2001.

9. Abraham T. (2001) IDDM: Intrusion Detection Using Data Mining Techniques. Technical Report DSTO-GD-0286, DSTO Electronics and Surveillance Research Laboratory

10. Valdes A. and Skinner K. (2000) Adaptive, model based monitoring for cyber attack detection. In Recent Advances on Intrusion Detection, pp 80-93, France, Springer Verlag

11. Ning P., Cui Y., Reeves D. S., Constructing Attack Scenarios through Correlation of Intrusion Alerts, Proc. ACM Computer and Communications Security Conf., 2002.

12. Ning P., Xu D., Learning Attack Strategies from Intrusion Alerts, Proc. ACM Computer and Communications Security Conf., 2003.

13. Cuppens F. and Miege A., Alert Correlation in a Cooperative Intrusion Detection Framework, Proc. IEEE Symposium on Security and Privacy, May 2002.

14. A. Singhal, "ANSWER: Network Monitoring using Object Oriented Rules" (with G. Weiss and J. Ros), Proceedings of the Tenth Conference on Innovative Application of Artificial Intelligence, Madison, Wisconsin, July 1998

15. Singhal A., "Design of GEMS Data Warehouse for AT&T Business Services", Proceedings of AT&T Software Architecture Symposium, Somerset, NJ, March 2000

16. Singhal A., "Design of Data Warehouse for Network/Web Services", Proceedings of Conference on Information and Knowledge Management (CIKM), November 2004. .

17. DARPA, DARPA 1998 Intrusion Detection Evaluation, **http://ideval.ll.mit.edu/1998_index.html**

18. SNORT, SNORT Intrusion Detection System, **http://www.snort.org**

19. RealSecure IDS, **http://www.iss.net**

20. KDD Cup 1999, **http://www.kdd.ics.uci.edu/databases/kddcup99/task.html**

21. GraphViz, Graph layout and drawing software, **http://www.research.att.com/sw/tools/graphviz**

22. X. Qin and W. Lee , "Statistical causality analysis of INFOSEC alert data", In Proceedings of 6th International Symposium on Recent Advances in Intrusion Detection (RAID 2003), September 2003.

23. X. Qin and W. Lee , "Discovering novel attack strategies from INFOSEC alerts", In Proceedings of the 9th European Symposium on Research in Computer Security (ESORICS 2004), September 2004.
24. Steven Noel, Eric Robertson, Sushil Jajodia, "Correlating Intrusion Events and Building Attack Scenarios through Attack Graph Distances," in *Proceedings of the 20th Annual Computer Security Applications Conference*, Tucson, Arizona, December 2004.

Chapter 6

MINDS: ARCHITECTURE & DESIGN

Varun Chandola, Eric Eilertson, Levent Ertoz, Gyorgy Simon and
Vipin Kumar
Department of Computer Science
University of Minnesota
{chandola,eric,ertoz,gsimon,kumar}@cs.umn.edu

Abstract   This chapter provides an overview of the Minnesota Intrusion Detection System (MINDS), which uses a suite of data mining based
algorithms to address different aspects of cyber security. The various components of MINDS such as the scan detector, anomaly detector and the profiling module detect different types of attacks
and intrusions on a computer network. The scan detector aims
at detecting scans which are the percusors to any network attack.
The anomaly detection algorithm is very effective in detecting behavioral anomalies in the network traffic which typically translate to malicious activities such as denial-of-service (DoS) traffic,
worms, policy violations and inside abuse. The profiling module
helps a network analyst to understand the characteristics of the
network traffic and detect any deviations from the normal profile.
Our analysis shows that the intrusions detected by MINDS are complementary to those of traditional signature based systems, such as
SNORT, which implies that they both can be combined to increase
overall attack coverage. MINDS has shown great operational success in detecting network intrusions in two live deployments at
the University of Minnesota and as a part of the Interrogator architecture at the US Army Research Lab - Center for Intrusion
Monitoring and Protection (ARL-CIMP).

Keywords: network intrusion detection, anomaly detection, summarization,
profiling, scan detection

The conventional approach to securing computer systems against cyber threats is to design mechanisms such as firewalls, authentication tools, and virtual private networks that create a protective shield. However, these mechanisms almost always have vulnerabilities. They cannot ward off attacks that are continually being adapted to exploit system weaknesses, which are often caused by careless design and implementation flaws. This has created the need for intrusion detection [6], security technology that complements conventional security approaches by monitoring systems and identifying computer attacks.

Traditional intrusion detection methods are based on human experts' extensive knowledge of attack signatures which are character strings in a messageŠs payload that indicate malicious content. Signatures have several limitations. They cannot detect novel attacks, because someone must manually revise the signature database beforehand for each new type of intrusion discovered. Once someone discovers a new attack and develops its signature, deploying that signature is often delayed. These limitations have led to an increasing interest in intrusion detection techniques based on data mining [12, 22, 2].

This chapter provides an overview of the *Minnesota Intrusion Detection System* (MINDS[1]) which is a suite of different data mining based techniques to address different aspects of cyber security. In Section 1 we will discuss the overall architecture of MINDS. In the subsequent sections we will briefly discuss the different components of MINDS which aid in intrusion detection using various data mining approaches.

1.   MINDS - **Minnesota INtrusion Detection System**

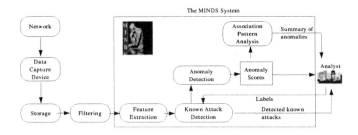

Figure 6.1. The Minnesota Intrusion Detection System (MINDS)

Figure 6.1 provides an overall architecture of the MINDS. The MINDS suite contains various modules for collecting and analyzing massive amounts of network traffic. Typical analyses include behavioral anomaly detection, summarization, scan detection and profiling. Additionally, the system has modules for feature extraction and filtering out attacks for which good signatures have been learnt [8]. Each of these modules will be individually described in the subsequent sections. Independently, each of these modules provides key insights into the network. When combined, which MINDS does automatically, these modules have a multiplicative affect on analysis. As shown in the figure, MINDS system is involves a network analyst who provides feedback to each of the modules based on their performance to fine tune them for more accurate analysis.

While the anomaly detection and scan detection modules aim at detecting actual attacks and other abnormal activities in the network traffic, the profiling module detects the dominant modes of traffic to provide an effective profile of the network to the analyst. The summarization module aims at providing a concise representation of the network traffic and is typically applied to the output of the anomaly detection module to allow the analyst to investigate the anomalous traffic in very few screen-shots.

The various modules operate on the network data in the NetFlow format by converting the raw network traffic using the *flow-tools* library [2]. Data in NetFlow format is a collection of records, where each record corresponds to a unidirectional flow of packets within a session. Thus each session (also referred to as a connection) between two hosts comprises of two flows in opposite directions. These records are highly compact containing summary information extracted primarily from the packet headers. This information includes source IP, source port, destination IP, destination port, number of packets, number of bytes and timestamp. Various modules extract more features from these basic features and apply data mining algorithms on the data set defined over the set of basic as well as derived features.

MINDS is deployed at the University of Minnesota, where several hundred million network flows are recorded from a network of more than 40,000 computers every day. MINDS is also part of the Interrogator [15] architecture at the US Army Research Lab - Center for Intrusion Monitoring and Protection (ARL-CIMP), where analysts collect and analyze network traffic from dozens of Department of Defense sites [7]. MINDS is enjoying great operational success at both sites, routinely detecting brand new attacks that signature-based systems could not have found. Additionally, it often discovers rogue communication channels and the exfiltration of data that other widely used tools such as SNORT [19] have had difficulty identifying.

2.   Anomaly Detection

Anomaly detection approaches build models of normal data and detect deviations from the normal model in observed data. Anomaly detection applied to intrusion detection and computer security has been an active area of research since it was originally proposed by Denning [6]. Anomaly detection algorithms have the advantage that they can detect emerging threats and attacks (which do not have signatures or labeled data corresponding to them) as deviations from normal usage. Moreover, unlike misuse detection schemes (which build classification models using labeled data and then classify an observation as normal or attack), anomaly detection algorithms do not require an explicitly labeled training data set, which is very desirable, as labeled data is difficult to obtain in a real network setting.

The MINDS anomaly detection module is a local outlier detection technique based on the *local outlier factor* (LOF) algorithm [3]. The LOF algorithm is effective in detecting outliers in data which has regions of varying densities (such as network data) and has been found to provide competitive performance for network traffic analysis[13].

The input to the anomaly detection algorithm is NetFlow data as described in the previous section. The algorithm extracts 8 *derived features* for each flow [8]. Figure 6.2 lists the set of features which are used to represent a network flow in the anomaly detection algorithm. Note that all of these features are either present in the NetFlow data or can be extracted from it without requiring to look at the packet contents.

Applying the LOF algorithm to network data involves computation of similarity between a pair of flows that contain a combination of categorical and numerical features. The anomaly detection algorithm uses a novel data-driven technique for calculating the distance between points in a high-dimensional space. Notably, this technique enables meaningful calculation of the similarity between records containing a mixture of categorical and numerical features shown in Figure 6.2.

Basic
Source IP
Source Port
Destination IP
Destination Port
Protocol
Duration
Packets Sent
Bytes per Packet Sent

Derived (Time-window Based)	
count-dest	Number of flows to unique destination IP addresses inside the network in the last T seconds from the same source
count-src	Number of flows from unique source IP addresses inside the network in the last T seconds to the same destination
count-serv-src	Number of flows from the source IP to the same destination port in the last T seconds
count-serv-dest	Number of flows to the destination IP address using same source port in the last T seconds

Derived (Connection Based)	
count-dest-conn	Number of flows to unique destination IP addresses inside the network in the last N flows from the same source
count-src-conn	Number of flows from unique source IP addresses inside the network in the last N flows to the same destination
count-serv-src-conn	Number of flows from the source IP to the same destination port in the last N flows
count-serv-dest-conn	Number of flows to the destination IP address using same source port in the last N flows

Figure 6.2. The set of features used by the MINDS anomaly detection algorithm

LOF requires the neighborhood around all data points be constructed. This involves calculating pairwise distances between all data points, which is an $O(n^2)$ process, which makes it computationally infeasible for a large number of data points. To address this problem, we sample a training set from the data and compare all data points to this small set, which reduces the complexity to $O(n * m)$ where n is the size of the data and m is the size of the sample. Apart from achieving computational efficiency, sampling also improves the quality of the anomaly detector output. The normal flows are very frequent and the anomalous flows are rare in the actual data. Hence the training data (which is drawn uniformly from the actual data) is more likely to contain several similar normal flows and far less likely to contain a substantial number of similar anomalous flows. Thus an anomalous flow will be unable to find similar anomalous neighbors in the training data and will have a high LOF score. The normal flows on the other hand will find enough similar normal flows in the training data and will have a low LOF score.

Thus the MINDS anomaly detection algorithm takes as input a set of network flows[3] and extracts a random sample as the training set. For each flow in the input data, it then computes its nearest neighbors in the training set. Using the nearest neighbor set it then computes the LOF score (referred to as the *Anomaly Score*) for that particular flow. The flows are then sorted based on their anomaly scores and presented to the analyst in a format described in the next section.

Output of Anomaly Detection Algorithm: The output of the MINDS anomaly detector is in plain text format with each input flow described in a single line. The flows are sorted according to their anomaly scores such that the top flow corresponds to the most anomalous flow (and hence most interesting for the analyst) according to the algorithm. For each flow, its anomaly score and the basic features describing that flow are displayed. Additionally, the contribution of each feature towards the anomaly score is also shown. The contribution of a particular feature signifies how different that flow was from its neighbors in that feature. This allows the analyst to understand the cause of the anomaly in terms of these features.

Table 6.1 is a screen-shot of the output generated by the MINDS anomaly detector from its live operation at the University of Minnesota. This output is for January 25, 2003 data which is one day after the Slammer worm hit the Internet. All the top 18 flows shown in Table 6.1 actually correspond to the worm related traffic generated by an external host to different U of M machines on destination port 1434 (which corresponds to the Slammer worm). The first entry in each line denotes the anomaly score of that

score	src IP	sPort	dst IP	dPort	proto	pkts	bytes
20826.69	128.171.X.62	1042	160.94.X.101	1434	tcp	[0,2)	[387,1264)
20344.83	128.171.X.62	1042	160.94.X.110	1434	tcp	[0,2)	[387,1264)
19295.82	128.171.X.62	1042	160.94.X.79	1434	tcp	[0,2)	[387,1264)
18717.1	128.171.X.62	1042	160.94.X.47	1434	tcp	[0,2)	[387,1264)
18147.16	128.171.X.62	1042	160.94.X.183	1434	tcp	[0,2)	[387,1264)
17484.13	128.171.X.62	1042	160.94.X.101	1434	tcp	[0,2)	[387,1264)
16715.61	128.171.X.62	1042	160.94.X.166	1434	tcp	[0,2)	[387,1264)
15973.26	128.171.X.62	1042	160.94.X.102	1434	tcp	[0,2)	[387,1264)
13084.25	128.171.X.62	1042	160.94.X.54	1434	tcp	[0,2)	[387,1264)
12797.73	128.171.X.62	1042	160.94.X.189	1434	tcp	[0,2)	[387,1264)
12428.45	128.171.X.62	1042	160.94.X.247	1434	tcp	[0,2)	[387,1264)
11245.21	128.171.X.62	1042	160.94.X.58	1434	tcp	[0,2)	[387,1264)
9327.98	128.171.X.62	1042	160.94.X.135	1434	tcp	[0,2)	[387,1264)
7468.52	128.171.X.62	1042	160.94.X.91	1434	tcp	[0,2)	[387,1264)
5489.69	128.171.X.62	1042	160.94.X.30	1434	tcp	[0,2)	[387,1264)
5070.5	128.171.X.62	1042	160.94.X.233	1434	tcp	[0,2)	[387,1264)
4558.72	128.171.X.62	1042	160.94.X.1	1434	tcp	[0,2)	[387,1264)
4225.09	128.171.X.62	1042	160.94.X.143	1434	tcp	[0,2)	[387,1264)

Table 6.1. Screen-shot of MINDS anomaly detection algorithm output for UofM data for January 25, 2003. The third octet of the IPs is anonymized for privacy preservation.

flow. The very high anomaly score for the top flows(the normal flows are assigned a score close to 1), illustrates the strength of the anomaly detection module in separating the anomalous traffic from the normal. Entries 2–7 show the basic features for each flow while the last entry lists all the features which had a significant contribution to the anomaly score. Thus we observe that the anomaly detector detects all worm related traffic as the top anomalies. A contribution vector for each of the flow (not shown in the figure due to lack of space) signifies that these anomalies were caused due to the feature – count_src_conn. The anomaly due to this particular feature translates to the fact that the external source was talking to an abnormally high number of inside hosts during a window of certain number of connections.

Table 6.2 shows another output screen-shot from the University of Minnesota network traffic for January 26, 2003 data (48 hours after the Slammer worm hit the Internet). By this time, the effect of the worm attack was reduced due to preventive measures taken by the network administrators. Table 6.2 shows the top 19 anomalous flows as ranked by the anomaly detector. Thus while most of the top anomalous flows still correspond to the worm traffic originating from an external host to different U of M machines on destination port 1434, there are two other type of anomalous flows which are highly ranked by the anomaly detector

1 Anomalous flows that correspond to a ping scan by an external host (Bold rows in Table 6.2)

2 Anomalous flows corresponding to U of M machines connecting to *half-life* game servers (Italicized rows in Table 6.2)

3.   Summarization

The ability to summarize large amounts of network traffic can be highly valuable for network security analysts who must often deal with large amounts of data. For example, when analysts use the MINDS anomaly detection algorithm to score several million network flows in a typical window of data, several hundred highly ranked flows might require attention. But due to the limited time available, analysts often can look only at the first few pages of results covering the top few dozen most anomalous flows. A careful look at the tables 6.1 and 6.2 shows that many of the anomalous flows are almost identical. If these similar flows can be condensed into a single line, it will enable the analyst to analyze a much larger set of anomalous flows. For example, the top 19 anom-

score	src IP	sPort	dst IP	dPort	proto	pkts	bytes
37674.69	63.150.X.253	1161	128.101.X.29	1434	tcp	[0,2)	[0,1829)
26676.62	63.150.X.253	1161	160.94.X.134	1434	tcp	[0,2)	[0,1829)
24323.55	63.150.X.253	1161	128.101.X.185	1434	tcp	[0,2)	[0,1829)
21169.49	63.150.X.253	1161	160.94.X.71	1434	tcp	[0,2)	[0,1829)
19525.31	63.150.X.253	1161	160.94.X.19	1434	tcp	[0,2)	[0,1829)
19235.39	63.150.X.253	1161	160.94.X.80	1434	tcp	[0,2)	[0,1829)
17679.1	63.150.X.253	1161	160.94.X.220	1434	tcp	[0,2)	[0,1829)
8183.58	63.150.X.253	1161	128.101.X.108	1434	tcp	[0,2)	[0,1829)
7142.98	63.150.X.253	1161	128.101.X.223	1434	tcp	[0,2)	[0,1829)
5139.01	63.150.X.253	1161	128.101.X.142	1434	tcp	[0,2)	[0,1829)
4048.49	**142.150.X.101**	**0**	**128.101.X.127**	**2048**	**icmp**	**[2,4)**	**[0,1829)**
4008.35	*200.250.Z.20*	*27016*	*128.101.X.116*	*4629*	*tcp*	*[2,4)*	*[0,1829)*
3657.23	*202.175.Z.237*	*27016*	*128.101.X.116*	*4148*	*tcp*	*[2,4)*	*[0,1829)*
2693.88	**142.150.X.101**	**0**	**128.101.X.168**	**2048**	**icmp**	**[2,4)**	**[0,1829)**
2444.16	**142.150.X.236**	**0**	**128.101.X.240**	**2048**	**icmp**	**[2,4)**	**[0,1829)**
2385.42	**142.150.X.101**	**0**	**128.101.X.45**	**2048**	**icmp**	**[0,2)**	**[0,1829)**
2057.15	**142.150.X.101**	**0**	**128.101.X.161**	**2048**	**icmp**	**[0,2)**	**[0,1829)**
1919.54	**142.150.X.101**	**0**	**128.101.X.99**	**2048**	**icmp**	**[2,4)**	**[0,1829)**
1634.38	**142.150.X.101**	**0**	**128.101.X.219**	**2048**	**icmp**	**[2,4)**	**[0,1829)**

Table 6.2. Screen-shot of MINDS anomaly detection algorithm output for UofM data for January 26, 2003. The third octet of the IPs is anonymized for privacy preservation.

alous flows shown in Table 6.2 can be represented as a three line summary as shown in Table 6.3. The column *count* indicates the number of flows represented by a line. "∗∗∗" indicates that the set of flows represented by the line had several distinct values for this feature. We observe that every flow is represented in the summary. The first summary represents flows corresponding to the *slammer worm* traffic coming from a single external host and targeting several internal hosts. The second summary represents connections made to *half-life* game servers by an internal host. The third summary corresponds to *ping scans* by different external hosts. Thus an analyst gets a fairly informative picture in just three lines. In general, such summarization has the potential to reduce the size of the data by several orders of magnitude. This motivates the need

avg Score	cnt	src IP	sPort	dst IP	dPort	proto
15102	10	63.150.X.253	1161	∗∗∗	1434	tcp
3833	2	∗∗∗	27016	128.101.X.116	∗∗∗	tcp
3371	7	∗∗∗	0	∗∗∗	2048	icmp

Table 6.3.   A three line summary of the 32 anomalous flows in Table 6.2.

to summarize the network flows into a smaller but meaningful representation. We have formulated a methodology for summarizing information in a database of transactions with categorical features as an optimization problem [4]. We formulate the problem of summarization of transactions that contain categorical data, as a dual-optimization problem and characterize a good summary using two metrics – *compaction gain* and *information loss*. Compaction gain signifies the amount of reduction done in the transformation from the actual data to a summary. Information loss is defined as the total amount of information missing over all original data transactions in the summary. We have developed several heurisitic algorithms which use frequent itemsets from the association analysis domain [1] as the candidate set for individual summaries and select a subset of these frequent itemsets to represent the original set of transactions.

The MINDS summarization module [8] is one such heuristic-based algorithm based on the above optimization framework. The input to the summarization module is the set of network flows which are scored by the anomaly detector. The summarization algorithm first generates frequent itemsets from these network flows (treating each flow as a transaction). It then greedily searches for a subset of these frequent itemsets such that the information loss incurred by the flows in the resulting summary is minimal. The summarization algorithm is further extended in MINDS by incorporating the ranks associated with the flows (based on the anomaly score). The underlying idea is that the highly ranked flows should incur very little loss, while the low ranked flows can be summarized in a more lossy manner. Furthermore, summaries that represent many anomalous flows (high scores) but few normal flows (low scores) are preferred. This is a desirable feature for the network analysts while summarizing the anomalous flows.

The summarization algorithm enables the analyst to better understand the nature of cyberattacks as well as create new signature rules for intrusion detection systems. Specifically, the MINDS summarization component compresses the anomaly detection output into a compact representation, so analysts can investigate numerous anomalous activities in a single screen-shot. Table 6.4 illustrates a typical MINDS output after anomaly detection and summarization. Each line contains the average anomaly score, the number of anomalous and normal flows represented by the line, eight basic flow features, and the relative contribution of each basic and derived anomaly detection feature (not shown in the figure due to lack of space). For example, the second line in Table 6.4 represents a total of 150 connections, of which 138 are highly anomalous. From this summary, analysts can easily infer that this is a backscatter from a denial-of-service attack on a computer that is outside the network being examined. Note that if an analyst looks at any one of these flows individually, it will be hard to infer that the flow belongs to back scatter even if the anomaly score is available. Similarly, lines 7, 17, 18, 19 together represent a total of 215 anomalous and 13 normal flows that represent summaries of FTP scans of the U of M network by an external host (200.75.X.2). Line 10 is a summary of IDENT lookups, where a remote computer is trying to get the user name of an account on an internal machine. Such inference is hard to make from individual flows even if the anomaly detection module ranks them highly.

4.   Profiling Network Traffic Using Clustering

Clustering is a widely used data mining technique [10, 24] which groups similar items, to obtain meaningful groups/clusters of data items in a data set. These clusters represent the dominant modes

score	c_1	c_2	src IP	sPort	dst IP	dPort	proto	pkts	bytes
31.17	-	-	218.19.X.168	5002	134.84.X.129	4182	tcp	[5,6)	[0,2045)
3.04	138	12	64.156.X.74	***	xx.xx.xx.xx	***	xxx	[0,2)	[0,2045)
15.41	-	-	218.19.X.168	5002	134.84.X.129	4896	tcp	[5,6)	[0,2045)
14.44	-	-	134.84.X.129	4770	218.19.X.168	5002	tcp	[5,6)	[0,2045)
7.81	-	-	134.84.X.129	3890	218.19.X.168	5002	tcp	[5,6)	***
3.09	4	1	xx.xx.xx.xx	4729	xx.xx.xx.xx	***	tcp	***	***
2.41	64	8	xx.xx.xx.xx	***	200.75.X.2	***	xxx	***	[0,2045)
6.64	-	-	218.19.X.168	5002	134.84.X.129	3676	tcp	[5,6)	[0,2045)
5.6	-	-	218.19.X.168	5002	134.84.X.129	4626	tcp	[5,6)	[0,2045)
2.7	12	0	xx.xx.xx.xx	***	xx.xx.xx.xx	113	tcp	[0,2)	[0,2045)
4.39	-	-	218.19.X.168	5002	134.84.X.129	4571	tcp	[5,6)	[0,2045)
4.34	-	-	218.19.X.168	5002	134.84.X.129	4572	tcp	[5,6)	[0,2045)
4.07	8	0	160.94.X.114	51827	64.8.X.60	119	tcp	[483,-)	[8424,-)
3.49	-	-	218.19.X.168	5002	134.84.X.129	4525	tcp	[5,6)	[0,2045)
3.48	-	-	218.19.X.168	5002	134.84.X.129	4524	tcp	[5,6)	[0,2045)
3.34	-	-	218.19.X.168	5002	134.84.X.129	4159	tcp	[5,6)	[0,2045)
2.46	51	0	200.75.X.2	***	xx.xx.xx.xx	21	tcp	***	[0,2045)
2.37	42	5	xx.xx.xx.xx	21	200.75.X.2	***	tcp	***	[0,2045)
2.45	58	0	200.75.X.2	***	xx.xx.xx.xx	21	tcp	***	[0,2045)

Table 6.4. Output of the MINDS summarization module. Each line contains an anomaly score, the number of anomalous and normal flows that the line represents, and several other pieces of information that help the analyst get a quick picture.

of behavior of the data objects determined using a similarity measure. A data analyst can get a high level understanding of the characteristics of the data set by analyzing the clusters. Clustering provides an effective solution to discover the expected and unexpected modes of behavior and to obtain a high level understanding of the network traffic.

The profiling module of MINDS essentially performs clustering, to find related network connections and thus discover dominant modes of behavior. MINDS uses the Shared Nearest Neighbor (SNN) clustering algorithm [9], which can find clusters of varying shapes, sizes and densities, even in the presence of noise and outliers. The algorithm can also handle data of high dimensionalities, and can automatically determine the number of clusters. Thus SNN is well-suited for network data. SNN is highly computationally intensive — of the order $O(n^2)$, where n is the number of network connections. We have developed a parallel formulation of the SNN clustering algorithm for behavior modeling, making it feasible to analyze massive amounts of network data.

An experiment we ran on a real network illustrates this approach as well as the computational power required to run SNN clustering on network data at a DoD site [7]. The data consisted of 850,000 connections collected over one hour. On a 16-CPU cluster, the SNN algorithm took 10 hours to run and required 100 Mbytes of memory at each node to calculate distances between points. The final clustering step required 500 Mbytes of memory at one node. The algorithm produced 3,135 clusters ranging in size from 10 to 500 records. Most large clusters correspond to normal behavior modes, such as virtual private network traffic. However, several smaller clusters correspond to deviant behavior modes that highlight misconfigured computers, insider abuse, and policy violations that are difficult to detect by manual inspection of network traffic.

Table 6.5 shows three such clusters obtained from this experiment. Cluster in Table 6.5(a) represents connections from inside machines to a site called GoToMyPC.com, which allows users (or attackers) to control desktops remotely. This is a policy violation in the organization for which this data was being analyzed. Cluster in Table 6.5(b) represents mysterious *ping* and SNMP traffic where a mis-configured internal machine is subjected to SNMP surveillance. Cluster in Table 6.5(c) represents traffic involving suspicious repeated *ftp* sessions. In this case, further investigations revealed that a mis-configured internal machine was trying

(a) Cluster representing connections to `GoToMyPC.com`

Duration	sIP	sPort	dIP	dPort	Pro	Pkt	Bytes
0:00:00	A	4125	B	8200	tcp	5	248
0:00:03	A	4127	B	8200	tcp	5	248
0:00:00	A	4138	B	8200	tcp	5	248
0:00:00	A	4141	B	8200	tcp	5	248
0:00:00	A	4143	B	8200	tcp	5	248
0:00:01	A	4149	B	8200	tcp	5	248
0:00:00	A	4163	B	8200	tcp	5	248
0:00:01	A	4172	B	8200	tcp	5	248
0:00:00	A	4173	B	8200	tcp	5	248
0:00:00	A	4178	B	8200	tcp	5	248

(b) Clusters representing mis-configured computers subjected to SNMP surveillance

Duration	sIP	sPort	dIP	dPort	Pro	Pkt	Bytes
0:00:00	A	1176	B	161	udp	1	95
0:00:00	A	-1	B	-1	icmp	1	84
0:00:00	A	1514	B	161	udp	1	95
0:00:00	A	-1	B	-1	icmp	1	84
0:00:00	A	-1	B	-1	icmp	1	84
0:00:00	A	-1	B	-1	icmp	1	84
0:00:00	A	-1	B	-1	icmp	1	84
0:00:00	A	3013	B	161	udp	1	95
0:00:00	A	-1	B	-1	icmp	1	84
0:00:00	A	3329	B	161	udp	1	95

(c) Cluster representing a mis-configured computer trying to contact Microsoft

Duration	sIP	sPort	dIP	dPort	Pro	Pkt	Bytes
0:00:00	A	3004	B	21	tcp	7	318
0:00:00	A	3007	B	21	tcp	7	318
0:00:00	A	3008	B	21	tcp	7	318
0:00:00	A	3011	B	21	tcp	7	318
0:00:00	A	3013	B	21	tcp	7	318
0:00:00	A	3015	B	21	tcp	7	318

Table 6.5.   Clusters obtained from network traffic at a US Army Fort

to contact Microsoft. Such clusters give analysts information they can act on immediately and can help them understand their network traffic behavior.

Table 6.6 shows a sample of interesting clusters obtained by performing a similar experiment on a sample of 7500 network flows sampled from the University of Minnesota network data. The first cluster (Table 6.6(a)) represent Kazaa (P2P) traffic between a U of M machine and different external P2P clients. Since Kazaa usage is not allowed in the university, this cluster brings forth an anomalous profile for the network analyst to investigate. Cluster in Table 6.6(b) represents traffic involving bulk data transfers between internal and external hosts; i.e. this cluster covers traffic in which the number of packets and bytes are much larger than the normal values for the involved IPs and ports. Cluster in Table 6.6(c) represents traffic between different U of M hosts and *Hotmail* servers (characterized by the port 1863). Cluster in Table 6.6(d) represents *ftp* traffic in which the data transferred is low. This cluster has different machines connecting to different *ftp* servers all of which are transferring much lower amount of data than the usual values for *ftp* traffic. A key observation to be made is that the clustering algorithm automatically determines the dimensions of interest in different clusters. In cluster of Table 6.6(a), the protocol, source port and the number of bytes are similar. In cluster of Table 6.6(b) the only common characteristic is large number of bytes. The common characteristics in cluster of Table 6.6(c) are the protocol and the source port. In cluster of Table 6.6(d) the common features are the protocol, source port and the low number of packets transferred.

5. Scan Detection

A precursor to many attacks on networks is often a reconnaissance operation, more commonly referred to as a scan. Identifying what attackers are scanning for can alert a system administrator or security analyst to what services or types of computers are being targeted. Knowing what services are being targeted before an attack allows an administrator to take preventative measures to protect the resources e.g. installing patches, firewalling services from the outside, or removing services on machines which do not need to be running them.

Given its importance, the problem of scan detection has been given a lot of attention by a large number of researchers in the

(a) Cluster representing Kazaa traffic between a UofM host and external machines

Duration	sIP	sPort	dIP	dPort	Pro	Pkt	Bytes
0:14:44	A_1	3531	B_1	3015	tcp	20	857
0:14:54	A_1	3531	B_2	4184	tcp	19	804
0:14:17	A_1	3531	B_3	10272	tcp	17	701
0:17:00	A_1	3531	B_4	4238	tcp	20	835
0:13:33	A_1	3531	B_5	2008	tcp	15	620

(b) Cluster representing bulk data transfer between different hosts

Duration	sIP	sPort	dIP	dPort	Pro	Pkt	Bytes
0:31:07	A_1	2819	B_1	4242	tcp	3154	129k
0:20:24	A_2	5100	B_2	1224	tcp	2196	121k
0:18:42	A_3	6881	B_3	1594	tcp	3200	4399k
0:15:08	A_4	4670	B_4	21	tcp	2571	3330k
0:10:20	A_5	27568	B_5	63144	tcp	2842	113k
0:09:00	A_6	6881	B_6	5371	tcp	2677	115k

(c) Cluster representing traffic between U of M hosts and *Hotmail* servers

Duration	sIP	sPort	dIP	dPort	Pro	Pkt	Bytes
00:00:00	A_1	1863	B_1	3969	tcp	1	41
00:00:30	A_2	1863	B_2	1462	tcp	4	189
00:00:00	A_3	1863	B_3	3963	tcp	1	41
00:00:00	A_4	1863	B_4	4493	tcp	1	41
00:00:50	A_5	1863	B_5	1102	tcp	4	176

(d) Cluster representing FTP traffic with small payload

Duration	sIP	sPort	dIP	dPort	Pro	Pkt	Bytes
00:00:02	A_1	21	B_1	1280	tcp	13	1046
00:00:05	A_1	21	B_2	34781	tcp	18	1532
00:00:11	A_2	21	B_3	9305	tcp	13	1185
00:00:00	A_1	21	B_4	27408	tcp	2	144
00:00:00	A_1	21	B_5	45607	tcp	4	227

Table 6.6.   Four clusters obtained from University of Minnesota network traffic

network security community. Initial solutions simply counted the number of destination IPs that a source IP made connection attempts to on each destination port and declared every source IP a scanner whose count exceeded a threshold [19]. Many enhancements have been proposed recently [23, 11, 18, 14, 17, 16], but despite the vast amount of expert knowledge spent on these methods, current, state-of-the-art solutions still suffer from high percentage of false alarms or low ratio of scan detection. For example, a recently developed scheme by Jung [11] has better performance than many earlier methods, but its performance is dependent on the selection of the thresholds. If a high threshold is selected, TRW will report only very few false alarms, but its coverage will not be satisfactory. Decreasing the threshold will increase the coverage, but only at the cost of introducing false alarms. P2P traffic and backscatter have patterns that are similar to scans, as such traffic results in many unsuccessful connection attempts from the same source to several destinations. Hence such traffic leads to false alarms by many existing scan detection schemes.

MINDS uses a data-mining-based approach to scan detection. Here we present an overview of this scheme and show that an off-the-shelf classifier, Ripper [5], can achieve outstanding performance both in terms of missing only very few scanners and also in terms of very low false alarm rate. Additional details are available in [20, 21].

Methodology: Currently our solution is a batch-mode implementation that analyzes data in windows of 20 minutes. For each 20-minute observation period, we transform the NetFlow data into a **summary data** set. Figure 6.3 depicts this process. With our focus on incoming scans, each new **summary record** corresponds to a potential scanner—that is pair of external source IP and destination port (SIDP). For each SIDP, the summary record contains a set of features constructed from the raw netflows available during the observation window. Observation window size of 20 minutes is somewhat arbitrary. It needs to be large enough to generate features that have reliable values, but short enough so that the construction of summary records does not take too much time or memory.

Given a set of summary data records corresponding to an observation period, scan detection can be viewed as a classification problem [24] in which each SIDP, whose source IP is external to the network being observed, is labeled as scanner if it was found

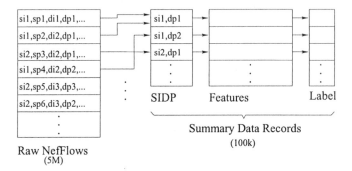

Figure 6.3. Transformation of raw netflow data in an observation window to the Summary Data Set.

scanning or `non-scanner` otherwise. This classification problem can be solved using predictive modeling techniques developed in the data mining and machine learning community if class labels (`scanner/non-scanner`) are available for a set of SIDPs that can be used as a training set.

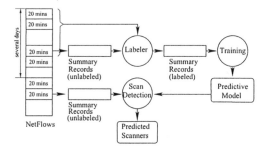

Figure 6.4. Scan Detection using an off-the-shelf classifier, Ripper.

Figure 6.4 depicts the overall paradigm. Each SIDP in the summary data set for an observation period (typically 20 minutes) is labeled by analyzing the behavior of the source IPs over a period of several days. The process involves two steps — (1) **Building a predictive model**: 20 minutes of `NetFlow` data is converted into unlabeled Summary Record format, which is labeled by the Labeler using several days of data. Predictive model is built on the labeled Summery Records. (2) **Scan Detection**: 20 minutes of data is converted into unlabeled Summary Record format. The predictive model is applied to it resulting in a list of predicted scanners.

The success of this method depends on (1) whether we can label the data accurately and (2) whether we have derived the right set of features that facilitate the extraction of knowledge. In the following sections, we will elaborate on these points.

Features: The key challenge in designing a data mining method for a concrete application is the necessity to integrate the expert knowledge into the method. A part of the knowledge integration is the derivation of the appropriate features. We make use of two types of expert knowledge. The first type of knowledge consists of a list of inactive IPs, a set of blocked ports and a list of P2P hosts in the network being monitored. This knowledge may be available to the security analyst or can be simply constructed by analyzing the network traffic data over a long period (several weeks or months). Since this information does not change rapidly, this analysis can be done relatively infrequently. The second type of knowledge captures the behavior of <source IP, destination port> (SIDP) pairs, based on the 20-minute observation window. Some of these features only use the second type of knowledge, and others use both types of knowledge.

Labeling the Data Set: The goal of labeling is to generate a data set that can be used as training data set for Ripper. Given a set of summarized records corresponding to 20-minutes of observation with unknown labels (unknown scanning statuses), the goal is to determine the actual labels with very high confidence. The problem of computing the labels is very similar to the problem of scan detection except that we have the flexibility to observe the behavior of an SIDP over a long period. This makes it possible to declare certain SIDPs as scanner or non-scanner with great confidence in many cases. For example, if a source IP s_ip makes a few failed connection attempts on a specific port in a short time window, it may be hard to declare it a scanner. But if the behavior of s_ip can be observed over a long period of time (e.g. few days), it can be labeled as non-scanner (if it mostly makes successful connections on this port) or scanner (if most of its connection attempts are to destinations that never offered service on this port). However, there will situations, in which the above analysis does not offer any clear-cut evidence one way or the other. In such cases, we label the SIDP as dontknow. For additional details on the labeling method, the reader is referred to [20].

Evaluation: For our experiments, we used real-world network trace data collected at the University of Minnesota between the 1st and the 22nd March, 2005. The University of Minnesota network consists of 5 class-B networks with many autonomous subnetworks. Most of the IP space is allocated, but many subnetworks have inactive IPs. We collected information about inactive IPs and P2P hosts over 22 days, and we used flows in 20 minute windows during 03/21/2005 (Mon.) and 03/22/2005 (Tue.) for constructing summary records for the experiments. We took samples of 20-minute duration every 3 hours starting at midnight on March 21. A model was built for each of the 13 periods and tested on the remaining 12 periods. This allowed us to reduce possible dependence on a certain time of the day, and performed our experiments on each sample.

Table 6.4 describes the traffic in terms of number of <source IP, destination port> (SIDP) combinations pertaining to scanning-, P2P-, normal- and backscatter traffic.

ID	Total	scan	p2p	normal	backscatter	dont-know
01	67522	3984	28911	6971	4431	23225
02	53333	5112	19442	9190	1544	18045
03	56242	5263	19485	8357	2521	20616
04	78713	5126	32573	10590	5115	25309
05	93557	4473	38980	12354	4053	33697
06	85343	3884	36358	10191	5383	29527
07	92284	4723	39738	10488	5876	31459
08	82941	4273	39372	8816	1074	29406
09	69894	4480	33077	5848	1371	25118
10	63621	4953	26859	4885	4993	21931
11	60703	5629	25436	4467	3241	21930
12	78608	4968	33783	7520	4535	27802
13	91741	4130	43473	6319	4187	33632

Table 6.7. The distribution of (source IP, destination ports) (SIDPs) over the various traffic types for each traffic sample produced by our labeling method

In our experimental evaluation, we provide comparison to TRW [11], as it is one of the state-of-the-art schemes. With the purpose of applying TRW for scanning worm containment, Weaver et al. [25] proposed a number of simplifications so that TRW can be implemented in hardware. One of the simplifications they applied—without significant loss of quality—is to perform the sequential hypothesis testing in logarithmic space. TRW then can be modeled as counting: a counter is assigned to each source IP and this counter is incremented upon a failed connection attempt and decremented upon a successful connection establishment.

Our implementation of TRW used in this paper for comparative evaluation draws from the above ideas. If the count exceeds a certain positive threshold, we declare the source to be scanner, and if the counter falls below a negative threshold, we declare the source to be normal.

The performance of a classifier is measured in terms of precision, recall and F-measure. For a contingency table of

	classified as Scanner	classified as not Scanner
actual Scanner	TP	FN
actual not Scanner	FP	TN

$$\text{precision} = \frac{TP}{TP + FP}$$

$$\text{recall} = \frac{TP}{TP + FN}$$

$$F - \text{measure} = \frac{2 * \text{prec} * \text{recall}}{\text{prec} + \text{recall}}.$$

Less formally, precision measures the percentage of scanning (source IP, destination port)-pairs (SIDPs) among the SIDPs that got declared scanners; recall measures the percentage of the actual scanners that were discovered; F-measure balances between precision and recall.

To obtain a high-level view of the performance of our scheme, we built a model on the 0321.0000 data set (ID 1) and tested it on the remaining 12 data sets. Figure 6.5 depicts the performance of our proposed scheme and that of TRW on the same data sets [4]. From left to right, the six box plots correspond to the precision, recall and F-measure of our proposed scheme and the precision, recall and F-measure of TRW. Each box plot has three lines corresponding (from top downwards) to the upper quartile, median and lower quartile of the performance values obtained over the 13 data sets. The whiskers depict the best and worst performance. One can see that not only does our proposed scheme outperform TRW by a wide margin, it is also more stable: the performance

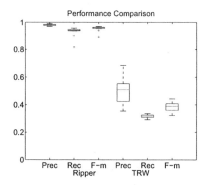

Figure 6.5.   Performance comparison between the proposed scheme and TRW.

varies less from data set to data set (the boxes in Figure 6.5 appear much smaller).

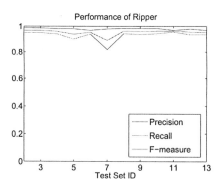

Figure 6.6.   The performance of the proposed scheme on the 13 data sets in terms of precision (topmost line), F-measure (middle line) and recall (bottom line).

Figure 6.6 shows the actual values of precision, recall and F-measure for the different data sets. The performance in terms of F-measure is consistently above 90% with very high precision, which is important, because high false alarm rates can rapidly deteriorate the usability of a system. The only jitter occurs on data set # 7 and it was caused by a single source IP that scanned a single destination host on 614(!) different destination ports meanwhile touching only 4 blocked ports. This source IP got misclassified as P2P, since touching many destination ports (on a number of IPs) is characteristic of P2P. This single misclassification introduced 614 false negatives (recall that we are classifying SIDPs not source IPs). The reason for the misclassification is that there were no vertical scanners in the training set — the highest number of destination ports scanned by a single source IP was 8, and this source IP touched over 47 destination IPs making it primarily a horizontal scanner.

6. Conclusion

MINDS is a suite of data mining algorithms which can be used as a tool by network analysts to defend the network against attacks and emerging cyber threats. The various components of MINDS such as the scan detector, anomaly detector and the profiling module detect different types of attacks and intrusions on a computer network. The scan detector aims at detecting scans which are the percusors to any network attack. The anomaly detection algorithm is very effective in detecting behavioral anomalies in the network traffic which typically translate to malicious activities such as *dos* traffic, worms, policy violations and inside abuse. The profiling module helps a network analyst to understand the characteristics of the network traffic and detect any deviations from the normal profile. Our analysis shows that the intrusions detected by MINDS are complementary to those of traditional signature based systems, such as SNORT, which implies that they both can be combined to increase overall attack coverage. MINDS has shown great operational success in detecting network intrusions in two live deployments at the University of Minnesota and as a part of the Interrogator [15] architecture at the US Army Research Lab - Center for Intrusion Monitoring and Protection (ARL-CIMP).

7. Acknowledgements

This work is supported by ARDA grant AR/F30602-03-C-0243, NSF grants IIS-0308264 and ACI-0325949, and the US Army High Performance Computing Research Center under contract DAAD 19-01-2-0014. The research reported in this article was performed in collaboration with Paul Dokas, Yongdae Kim, Aleksandar Lazarevic, Haiyang Liu, Mark Shaneck, Jaideep Srivastava, Michael Steinbach, Pang-Ning Tan, and Zhi-li Zhang. Access to computing facilities was provided by the AHPCRC and the Minnesota Super-computing Institute.

Notes

1. www.cs.umn.edu/research/minds
2. www.splintered.net/sw/flow-tools
3. Typically, for a large sized network such as the University of Minnesota, data for a 10 minute long window is analyzed together
4. The authors of TRW recommend a threshold of 4. In our experiments, we found, that TRW can achieve better performance (in terms of F-measure) when we set the threshold to 2, this is the threshold that was used in Figure 6.5, too.

References

[1] Rakesh Agrawal, Tomasz Imieliski, and Arun Swami. Mining association rules between sets of items in large databases. In *Proceedings of the 1993 ACM SIGMOD international conference on Management of data*, pages 207–216. ACM Press, 1993.

[2] Daniel Barbara and Sushil Jajodia, editors. *Applications of Data Mining in Computer Security*. Kluwer Academic Publishers, Norwell, MA, USA, 2002.

[3] Markus M. Breunig, Hans-Peter Kriegel, Raymond T. Ng, and J Sander. Lof: identifying density-based local outliers. In *Proceedings of the 2000 ACM SIGMOD international conference on Management of data*, pages 93–104. ACM Press, 2000.

[4] Varun Chandola and Vipin Kumar. Summarization – compressing data into an informative representation. In *Fifth IEEE International Conference on Data Mining*, pages 98–105, Houston, TX, November 2005.

[5] William W. Cohen. Fast effective rule induction. In *International Conference on Machine Learning (ICML)*, 1995.

[6] Dorothy E. Denning. An intrusion-detection model. *IEEE Trans. Softw. Eng.*, 13(2):222–232, 1987.

[7] Eric Eilertson, Levent Ertoz, Vipin Kumar, and Kerry Long. Minds – a new approach to the information security process. In 24th *Army Science Conference*. US Army, 2004.

[8] Levent Ertoz, Eric Eilertson, Aleksander Lazarevic, Pang-Ning Tan, Vipin Kumar, Jaideep Srivastava, and Paul Dokas. MINDS - Minnesota Intrusion Detection System. In *Data Mining - Next Generation Challenges and Future Directions*. MIT Press, 2004.

[9] Levent Ertoz, Michael Steinbach, and Vipin Kumar. Finding clusters of different sizes, shapes, and densities in noisy, high dimensional data. In *Proceedings of 3rd SIAM International Conference on Data Mining*, May 2003.

[10] Anil K. Jain and Richard C. Dubes. *Algorithms for Clustering Data.* Prentice-Hall, Inc., 1988.

[11] Jaeyeon Jung, Vern Paxson, Arthur W. Berger, and Hari Balakrishnan. Fast portscan detection using sequential hypothesis testing. In *IEEE Symposium on Security and Privacy*, 2004.

[12] Vipin Kumar, Jaideep Srivastava, and Aleksander Lazarevic, editors. *Managing Cyber Threats–Issues, Approaches and Challenges.* Springer Verlag, May 2005.

[13] Aleksandar Lazarevic, Levent Ertoz, Vipin Kumar, Aysel Ozgur, and Jaideep Srivastava. A comparative study of anomaly detection schemes in network intrusion detection. In *SIAM Conference on Data Mining (SDM)*, 2003.

[14] C. Lickie and R. Kotagiri. A probabilistic approach to detecting network scans. In *Eighth IEEE Network Operations and Management*, 2002.

[15] Kerry Long. Catching the cyber-spy, arl's interrogator. In 24th *Army Science Conference*. US Army, 2004.

[16] V. Paxon. Bro: a system for detecting network intruders in real-time. In *Eighth IEEE Network Operators and Management Symposium (NOMS)*, 2002.

[17] Phillip A. Porras and Alfonso Valdes. Live traffic analysis of tcp/ip gateways. In *NDSS*, 1998.

[18] Seth Robertson, Eric V. Siegel, Matt Miller, and Salvatore J. Stolfo. Surveillance detection in high bandwidth environments. In *DARPA DISCEX III Conference*, 2003.

[19] Martin Roesch. Snort: Lightweight intrusion detection for networks. In *LISA*, pages 229–238, 1999.

[20] Gyorgy Simon, Hui Xiong, Eric Eilertson, and Vipin Kumar. Scan detection: A data mining approach. Technical Report AHPCRC 038, University of Minnesota – Twin Cities, 2005.

[21] Gyorgy Simon, Hui Xiong, Eric Eilertson, and Vipin Kumar. Scan detection: A data mining approach. In *Proceedings of SIAM Conference on Data Mining (SDM)*, 2006.

[22] Anoop Singhal and Sushil Jajodia. Data mining for intrusion detection. In *Data Mining and Knowledge Discovery Handbook*, pages 1225–1237. Springer, 2005.

[23] Stuart Staniford, James A. Hoagland, and Joseph M. McAlerney. Practical automated detection of stealthy portscans. *Journal of Computer Security*, 10(1/2):105–136, 2002.

[24] Pang-Ning Tan, Michael Steinbach, and Vipin Kumar. *Introduction to Data Mining*. Addison-Wesley, May 2005.

[25] Nicholas Weaver, Stuart Staniford, and Vern Paxson. Very fast containment of scanning worms. In *13th USENIX Security Symposium*, 2004.

Chapter 7

DISCOVERING NOVEL ATTACK STRATEGIES FROM INFOSEC ALERTS

Xinzhou Qin*
Cisco Systems, Inc.
210 West Tasman Dr., San Jose, CA
keqin@cisco.com

Wenke Lee
College of Computing, Georgia Institute of Technology
Atlanta, GA 30332
wenke@cc.gatech.edu

Abstract   Deploying a large number of information security (INFOSEC) systems can provide in-depth protection for systems and networks. However, the sheer number of security alerts output by security sensors can overwhelm security analysts and keep them from performing effective analysis and initiating timely response. Therefore, it is important to develop an advanced alert correlation system that can reduce alarm redundancy, intelligently correlate security alerts and detect attack strategies. Alert correlation is therefore a core component of a security management system.

Correlating security alerts and discovering attack strategies are important and challenging tasks for security analysts. Recently, there have been several proposed techniques to analyze attack scenarios from security alerts. However, most of these approaches depend on *a priori* and hard-coded domain knowledge that lead to their limited capabilities of detecting new attack strategies. In addition, these approaches focus more on the aggregation and analysis of raw security alerts, and build basic or low-level attack scenarios.

This paper focuses on discovering novel attack strategies via analysis of security alerts. Our integrated alert correlation system helps security administrator aggregate redundant alerts, filter out unrelated attacks, correlate security alerts and analyze attack scenarios.

Our integrated correlation system consists of three complementary correlation mechanisms based on two hypotheses of attack step relationship. The first

*The work was done when the author was at College of Computing at Georgia Institute of Technology.

hypothesis is that some attack steps are directly related because an earlier attack enables or positively affects the later one. We have developed a probabilistic-based correlation engine that incorporates domain knowledge to correlate alerts with direct causal relationship. The second hypothesis is that some related attack steps, even though they do not have obvious or direct (or known) relationship in terms of security and performance measures, still exhibit statistical and temporal patterns. For this category of relationship, we have developed two correlation engines to discover attack transition patterns based on statistical analysis and temporal pattern analysis, respectively. Based on the correlation results of these three correlation engines, we construct attack scenarios and conduct attack path analysis. The security analysts are presented with aggregated information on attack strategies from the integrated correlation system.

We evaluate our approaches using DARPA's Grand Challenge Problem (GCP) data sets. Our evaluation shows that our approach can effectively discover novel attack strategies, provide a quantitative analysis of attack scenarios and identify attack plans.

Keywords: Security alert correlation, intrusion detection, security management

1.   Introduction

Information security (INFOSEC) is a complex process with many challenging problems. As more security systems are developed, deploying a large scale of INFOSEC mechanisms, e.g., authentication systems, firewalls, intrusion detection systems (IDSs), antivirus software, network management and monitoring systems, can provide protection in depth for the IT infrastructure. INFOSEC sensors often output a large quantity of low-level or incomplete security alerts because there is a large number of network and system activities being monitored and multiple INFOSEC systems can each report some aspects of security events. The sheer quantity of alerts from these security systems and sensors can overwhelm security administrators and prevent them from performing comprehensive security analysis of the protected domains and initiating timely response.

From a security administrator's point of view, it is important to reduce the redundancy of alarms, intelligently integrate and analyze security alerts, construct attack scenarios (defined as a sequence of related attack steps) and present high-level aggregated information from multiple local-scale events. To address this issue, researchers and security product vendors have proposed *alert correlation*, a process to analyze and correlate security alerts to provide an aggregated information on the networks and systems under protection. Applying alert correlation techniques to identifying attack scenarios can also help forensic analysis, response and recovery, and even prediction of forthcoming attacks. Therefore, alert correlation is a core component in a security management system.

Recently there have been several proposals on alert correlation, including alert similarity measurement [52], probabilistic reasoning [19], clustering algorithms [14], pre- and post-condition matching of known attacks [39, 12, 7], and chronicles formalism approach [38]. Most of these proposed approaches have limited capabilities because they rely on various forms of predefined knowledge of attack conditions and consequences. They cannot recognize a correlation when an attack is new or the relationship between attacks is new. In other words, these approaches in principle are similar to *misuse detection* techniques, which use the "signatures" of known attacks to perform pattern matching and cannot detect new attacks. It is obvious that the number of possible correlations is very large, potentially a combinatorial of the number of known and new attacks. It is infeasible to know *a priori* and encode all possible matching conditions between attacks. To further complicate the matter, the more dangerous and intelligent adversaries will always invent new attacks and novel attack sequences. Therefore, we must develop significantly better alert correlation algorithms that can discover sophisticated and new attack sequences.

We have two motivations in our work. First, we want to develop an alert correlation system that can discover *new* attack strategies without relying solely on domain knowledge. Second, we want to incorporate more evidence or indicators from other non-security monitoring systems to correlate alerts and detect attack strategies. For example, we can incorporate alerts from network management systems (NMS) into the security alert correlation. Although alerts from NMS may not directly tell us what attacks are present, they provide us information on the state of protected domains.

This paper focuses on correlation techniques. Our main contribution in this paper is the design of an integrated correlation system to discover novel attack strategies from INFOSEC alerts. Our alert correlation mechanism integrates three different correlation methods based on two hypotheses of attack step relationships to discover and analyze relationships among alerts. *Bayesian-based correlation engine* [48] applies probabilistic reasoning to correlate alerts that have direct causal relationships according to some domain knowledge. This correlation mechanism is based on the hypothesis that some attack steps have direct relationship because prior attack step enables the later one. *Causal discovery theory-based correlation mechanism* performs alert correlation using statistical analysis of attack occurrences to identify the dependency between alerts. *Time series-based* correlation engine [46] conducts alert correlation using statistical test and investigating temporal relationship between alerts. These two statistical and temporal-based correlation mechanisms are based on the hypothesis that some attack steps have temporal or statistical patterns even though they may not have direct or obvious (or known) relationships in terms of security or performance measures. We integrate these three complementary correlation engines to perform alert analysis and correlation. We construct attack scenarios and conduct attack path analysis based on the output of three correlation engines. We evaluate and rank the overall likelihood of various attack paths and identify those with higher probabilities. The result of alert correlation is a set of candidate attack plans corresponding to the intrusions executed by the attacker. The outputs of this phase can be used for further analysis in the later phase, i.e., *attack plan recognition* [47].

We evaluate our methods using DARPA's Grand Challenge Problem (GCP) data sets [13]. The results show that our approach can successfully discover new attack strategies and provide a quantitative analysis method to analyze attack strategies.

The remainder of this paper is organized as follows. We describe our method of alert aggregation and prioritization in Section 2. We present our probabilistic-based correlation mechanism in Section 3. We describe our statistical-based alert correlation engine in Section 4. Causal discovery-based correlation engine is described in Section 5. In Section 6, we present our approach to integrate these three correlation engines and scenario analysis. In Section 7, we report the experiments and results on the GCP. Section 8 discusses the related work. We summarize the paper and point out some ongoing and future work in Section 9.

2.   Alert Aggregation and Prioritization

In this section, we describe two major components in our alert correlation system, i.e., alert aggregation and alert prioritization.

2.1   Alert Aggregation and Clustering

One of the issues with deploying multiple security devices is the large number of alerts output by the devices. The large volume of alerts make it very difficult for the security administrator to analyze attack events and handle alerts in a timely fashion. Therefore, the first step in alert analysis is alert aggregation and volume reduction.

In our approach, we use alert fusion and clustering techniques to reduce the redundancy of alerts while keeping the important information. Specifically, each alert has a number of attributes such as *time stamp*, *source IP*, *destination IP*, *port(s)*, *user name*, *process name*, *attack class*, and *sensor ID*, which are defined in a standard document named "Intrusion Detection Message Exchange Format (IDMEF)" drafted by IETF Intrusion Detection Working Group [21].

IDMEF has defined alert formats and attributes. IDMEF is intended to be a standard data format that intrusion detection systems can use to report alerts about suspicious events. A Document Type Definition (DTD) has been proposed to describe IDMEF data format by XML documents.

In IDMEF, three temporal attributes have been defined to be associated to an alert. *Detect-time* refers to the time that the attack occurs, *create-time* represents the time when the attack is detected and *analyzer-time* is the time when the alert is output by an IDS. *Create-time* and *analyzer-time* are fully dependant on the characteristics of the IDS. Therefore, we use *detect-time* attributes in our alert aggregation process. In other words, two alerts might be considered similar even though their *create-time* and *analyzer-time* are completely different.

In the IDMEF format, the structures of attributes source and target are similar. They can be described by a node, a user, a process and a service. A node might be identified by its IP address (typically by a network-based IDS) or by its host name (typically by a host-based IDS). Similarly, some IDSs provide service names or port numbers. We create and use two correspondence tables between host names and IP addresses, and between services and port numbers. For most alerts output by a host-based IDS, we specify that a similarity exists between alerts' source and target attributes if both their nodes, users, services and processes are similar. And for most network attacks, we compare the nodes and services.

Alert fusion has two phases, i.e., aggregation of alerts of the same IDS and aggregation of alerts of different sensors. Specifically, we first combine alerts that have the same attributes except time stamps. This step is intended to aggregate alerts that are output by the same IDS and are corresponding to the same attack but have a small delay, i.e., the time stamps of those alerts can be slightly different, e.g., two seconds apart. Second, based on the results of step 1, we aggregate alerts with the same attributes but are reported from different heterogeneous sensors. The alerts varied on time stamp are fused together if they are close enough to fall in a pre-defined time window.

Alert clustering is used to further group alerts after alert fusion. Based on various clustering algorithms, we can group alerts in different ways according to the *similarity* among alerts, (e.g., [52] and [30]). Currently, based on the results of alert fusion, we further group alerts that have same attributes except time stamps into one cluster. After this step, we have further reduced the redundancy of alerts.

DEFINITION 1 *A hyper alert is defined as a time ordered sequence of alerts that belong to the same cluster.*

For example, after alert clustering, we have a series of aggregated alert instances, $a_1, a_2...a_n$, in one cluster that have the same attributes along the time axis. We use hyper alert A to represent this sequence of alerts, i.e., $A = \{a_1, a_2, ..., a_n\}$.

2.2 Alert Verification and Prioritization

The next phase of alert processing is to verify and prioritize each hyper alert based on its success and relevance to the mission goals.

When a correlation engine receives false positives as input, the quality of correlation results can degrade significantly. Therefore, the reduction of false positive and irrelevant alerts is an important prerequisite to achieve a good correlation results.

The task of alert verification is to examine an alert and determine the success or failure of the corresponding attack. It aims to filter out the false positive alerts output by security sensors.

We apply evidence cross checking to identifying the false positive alert. In other words, we use alerts or evidence output by other security sensors to cross check the validity of an alert. In particular, for an alert generated by a security sensor (e.g., an IDS), we check if there are any similar alerts output by other security sensors or if there are any alerts or evidence corresponding to the impact of the attack. For example, when a network-based IDS output a buffer overflow alert targeting a specific process running on the target host, and if the host-based IDS installed on the target machine also generated an alert representing an abnormal running of that process or other abnormal activities (e.g., illegal file access) corresponding to the evidence of the attack impact, then we can enforce the validity of the buffer overflow alert.

Priorities are important to classify alerts and quickly discard information that is irrelevant or of less importance to a particular site. The alert prioritizing component has to take into account the security policy and the security requirements of the site where the correlation system is deployed. The objective is that, with the alert priority rank, security analyst can select important alerts as the target alerts for further correlation and analysis. Specifically, the priority score of an alert is computed based on the relevance of the alert to the configuration of the protected networks and hosts as well as the severity of the corresponding attack assessed by the security analyst. In practice, a correlation system uses the information from the impact analysis and the asset database to determine the importance of network services to the overall mission goals of the network.

Porras et al. proposed a more comprehensive mechanism of incident/alert rank computation model in a "mission-impact-based" correlation engine, named M-Correlator [45]. Since we focus on alert correlation and scenario analysis instead of alert priority ranking, and alert prioritization is just an intermediate step to facilitate further alert analysis, we adapted the priority computation model of M-Correlator with a simplified design.

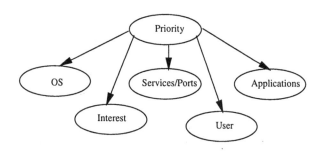

Figure 7.1. Alert Priority Computation Model

Figure 7.1 shows our priority computation model that is constructed based on Bayesian networks [43]. We use Bayesian inference to obtain a belief over states (hypotheses) of interests. A Bayesian network is usually represented as a directed acyclic graph (DAG) where each node represents a variable, and the directed edges represent the causal or dependent relationships among the variables. A conditional probability table (CPT) [43] is associated with each child node. It encodes the prior knowledge between the child node and its parent node. Specifically, an element of the CPT at a child node is defined by $CPT_{ij} = P(child\_state = j | parent\_state = i)$ [43]. The belief in hypotheses of the root is related to the belief propagation from its child nodes, and ultimately the evidence at the leaf nodes.

Specifically, in our priority computation model, the root represents the priority with two hypothesis states, i.e., "high" and "low". Each leaf node has three states. For node "Interest", its three states are "low", "medium" and "high". For other nodes, the three states are "matched", "unmatched" and "unknown". The computation result is a value in [0,1] where 1 is the highest priority score.

We denote e^k as the kth leaf node and H_i as the ith hypothesis of the root node. Given the evidence from the leaf nodes, assuming conditional independence with respect to each H_i, the belief in hypothesis at the root is: $P(H_i \mid e^1, e^2, \ldots, e^N) = \gamma P(H_i) \prod_{k=1}^{N} P(e^k | H_i)$, where $\gamma = [P(e^1, e^2, \ldots, e^N)]^{-1}$ and γ can be computed using the constraint $\sum_i P(H_i | e^1, e^2, \ldots, e^N) = 1$. For example, for the hyper alert of *FTP Globbing Buffer Overflow* attack, we get evidence [*high, matched, matched, unknown, unknown*] from the corresponding leaf nodes, i.e., Interest, OS, Services/Ports, Applications and User, respectively. As Figure 7.1 shows, the root node represents the priority of hyper alert. Assume that we have the prior probabilities for the hypotheses of the root, i.e., $P(Priority = high) = 0.8$ and $P(Priority = low) = 0.2$, and the following conditional probabilities as defined in the CPT at each leaf node, $P(Interest = high | Priority = high) = 0.70$, $P(Interest = high | Priority = low) = 0.10$, $P(OS = matched | Priority = high) = 0.75$, $P(OS = matched | Priority = low) = 0.20$, $P(Services = matched | Priority = high) = 0.70$, $P(Services = matched | Priority = low) = 0.30$, $P(Applications = unknown | Priority = high) = 0.15$, $P(Applications = unknown | Priority = low) = 0.15$, $P(User = unknown | Priority = high) = 0.10$, $P(User = unknown | Priority = low) = 0.10$, we then can get $\gamma = 226.3468$, therefore, $P(Priority = high | Interest = matched, OS = matched, Service = matched, Applications = matched, User = unknown) = 0.9959$. We regard this probability as the priority score of the alert. The current CPTs are predefined based on our experience and domain knowledge.

To calculate the priority of each hyper alert, we compare the dependencies of the corresponding attack represented by the hyper alert against the configurations of target networks and hosts. We have a knowledge base in which each hyper alert has been associated with a few fields that indicate its attacking OS, services/ports and applications. For the alert output from a host-based IDS, we will further check if the target user exists in the host configuration. The purpose of relevance check is that we can downgrade the importance of some alerts that are unrelated to the protected domains. For example, an attacker may launch an individual buffer overflow attack against a service blindly, without knowing if the service exists. It is quite possible that a signature-based IDS outputs the alert once the packet contents match the detection rules even though such service does not exist on the protected host. The relevance check on the alerts aims to downgrade the impact of such kind of alerts on further correlation analysis. The interest of the attack is assigned by the security analyst based on the nature of the attack and missions of the target hosts and services in the protected domain.

3.   Probabilistic-Based Alert Correlation

3.1   Motivation

In practice, we observe that when a host is compromised by an attacker, it usually becomes the target of further attacks or a stepping-stone for launching attacks against other systems. Therefore, the consequences of an attack on a compromised host can be used to reason about a possible matching with the goals of another attack. In a series of attacks where the attackers launch earlier attacks to prepare for later ones, there are usually strong connections between the consequences of the earlier attacks and the prerequisites of the later ones. If an earlier attack is to prepare for a later attack, the consequence of the earlier attack should at least partly satisfy the prerequisite of the later attack.

It is possible to address this type of correlation by defining pre- and post-conditions of individual attacks and applying condition matching. However, it is infeasible to enumerate and precisely encode all possible attack consequences and goals into pre- and post-conditions. In addition, in practice, an attacker does not have to perform early attacks to prepare for a later one, even though the later attack has certain prerequisites. For example, an attacker can launch an individual buffer overflow attack against a service blindly without knowing if the service exists or not. In other words, the prerequisite of an attack should not be mistaken for the necessary existence of an earlier attack. A hard-coded pre- and post-conditions matching approach cannot handle such cases.

Having the challenges in mind, we apply probabilistic reasoning to alert correlation by incorporating system indicators of attack consequences and prior knowledge of attack transitions. In this section, we discuss how to apply probabilistic reasoning to attack consequences and goals in order to discover the subtle relationships between attack steps in an attack scenario.

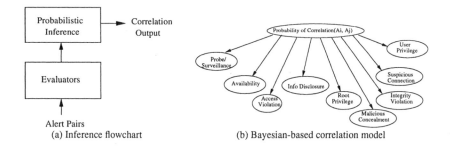

Figure 7.2. Probabilistic reasoning model

3.2   Model Description

Figure 7.2(a) shows the procedure of correlation inference. Given a stream of alerts, *evaluators* first analyze one or more features of alert pairs and output results as evidence to the *inference module*. The *inference module* combines the individual opinions expressed by the evaluators into a single assessment of the correlation by computing and propagating correlation beliefs within the inference network.

In our inference module, we use a Bayesian network [43] as our reasoning engine. Bayesian networks are usually used as a principle method to reason uncertainty and are capable of leveraging prior expert opinions with the learned information from data. A Bayesian network is usually represented as a directed acyclic graph (DAG) where each node represents a variable that has a certain set of states, and the directed edges represent the causal or dependent relationships among the variables. A Bayesian network consists of several parameters, i.e., prior probability of parent node's states (i.e., $P(parent\_state = i)$), and a set of conditional probability tables (CPT) associated with child nodes. CPT encodes the prior knowledge between child node and its parent node. Specifically, an entry of the CPT at a child node is defined by $CPT_{ij} = P(child\_state = j|parent\_state = i)$. We have more discussions on probability properties of a Bayesian network in Section 5.2.1.

Figure 7.2(b) shows the structure of our Bayesian inference model for pairwise correlation. Since we depend on domain knowledge to correlate directly related alert pairs, we design a one-level Bayesian network that is good enough to perform inference.

In the inference model, the root node represents the *hypothesis* that two attacks are correlated. Specifically, the root node has two hypothesis states, i.e., "high correlation" and "low correlation". Each child node represents a type of attack consequences on the host. The evaluator on each child node detects the condition matching between the consequences and the necessary conditions of the two alerts being correlated. The evaluation result on each leaf node is mapped to a state of the child node. Each child node has three states: "matched", "not matched" and "unknown". The state "unknown" handles the case that there is no need of condition matching, e.g., some attacks do not necessarily have any pre-conditions in order to be launched. The output of the inference engine represents the probability or confidence of the correlation between two alerts being analyzed (i.e., $P(correlation = high|evidence)$) based on the evidence (e.g., "matched" or "unmatched") provided by the leaf nodes.

The belief computation is conducted by propagating belief messages among leaf nodes and the root node. Specifically, we denote e^k as the kth leaf node and H_i as the ith hypothesis of the root node. Given the evidence from the leaf nodes, assuming conditional independence with respect to each H_i, the belief in hypothesis at the root is: $P(H_i \mid e^1, e^2, \ldots, e^N) = \gamma P(H_i) \prod_{k=1}^{N} P(e^k|H_i)$, where $\gamma = [P(e^1, e^2, \ldots, e^N)]^{-1}$ and γ can be computed using the constraint $\sum_i P(H_i|e^1, e^2, \ldots, e^N) = 1$ [43]. Since the belief computation can be performed incrementally instead of being delayed until all the evidence is collected, the Bayesian inference engine can also function on partial evidence, and the lack of evidence input from an evaluator does not require special treatment.

As Figure 7.2(b) shows, each leaf node represents an attack consequence on the attack victim.

When reasoning about the correlation between two alerts, we consider broad aspects of attack consequences, in particular, (1) *Probe or Surveillance*: information on system or network has been gained by an attacker, e.g., a probing attack can get information on open ports. (2) *Availability*: the system is out of service or the service is negatively affected by the attack, e.g., because of a DoS attack. (3) *Access Violation*: an illegal access to a file or data of a system. (4) *Information Disclosure*: the attacker exports (sensitive) data to external site. (5) *Root Privilege* has been obtained by an attacker, for example, by a buffer overflow attack. (6) *Malicious Concealment*: malicious binary codes have been installed on the system, e.g., a Trojan horse. (7) *Integrity Violation*: the file on a system has been modified or deleted, violating the security policy. (8) *Suspicious Connection*: a covert channel has been set up by the attack. (9) *User Privilege* has been obtained by the attacker.

For each attack, it may result in one or more of those impacts on the victim host or network. Each attack may also need some pre-conditions prepared by prior attack(s) in one or more above fields. Therefore, when correlating two alerts, we compare the causal alert candidate's consequences with effected alert's pre-conditions in each leaf nodes of Figure 7.2(b).

Table 7.1 shows the set of predicates that we defined to assess the consequences of attack. Each attack impact shown in Figure 7.2(b) has been associated with a set of predicates defined in Table 7.1.

Table 7.1. Predicates used in impact evaluation

FailService	DegradeService	FailProcess
DegradeProcess	ModifyData	DeleteData
GainUserPrivilege	GainRootPrivilege	GainServiceInfo
GainOSInfo	InstallMaliciousDaemon	InstallTrojan
SetupCovertChannel	FailCovertChannel	ExportData
GainFile	AccessSystem	LeakInformation

For example, predicates "FailService" and "DegradeService" represent the attack impacts on the availability of the target's service. The definition of predicates is a broad template and each predicate can be instantiated to a specific consequence instance according to information provided by alerts. For example, when a port scan alert is output, its corresponding impact instance is *GainServiceInfo.targetIP*. For another example, an attack may result in compromise of the *root privilege* and modification of the *password* file at a victim host. The corresponding attack consequence can be represented by {*GainRootPrivilege.targetIP, ModifiyData.passwordFile*}.

Each alert has also been defined a *pre-condition(s)* using the predicates shown in Table 7.1. Like the definition of impact of attack, *pre-condition(s)* of each alert can also be instantiated based on alert specific attributes. Each alert can provide the necessary information from its attributes, such as *source IP, target IP, attack class*.

Correlating two alerts includes the following steps. First, each alert first initializes its corresponding *pre-condition* and *impact* fields. Second, alert pairs are checked to see if they comply with certain constraints, e.g., an implicit temporal constraint between these two alerts is that alert A_i occurs before alert A_j. Third, evaluations are conducted by comparing the *causal* alert's impacts and *effected* alert's pre-conditions on each of the leaf nodes as shown in Figure 7.2. Fourth, results of evaluations are mapped to the states of leaf nodes, i.e., "matched", "unmatched" and "unknown". Finally, an overall probability computation is conducted based on the state evidence of each leaf node.

For example, alert *portscan* has a consequence defined as *GainServiceInfo.targetIP* that is associated with attack consequence *Probe or Surveillance* as shown in Figure 7.2(b). Alert *imap buffer overflow* has a pre-condition as *GainServiceInfo.targetIP*, where predicate "GainServiceInfo" is associated with attack consequence *Probe/Surveillance* shown in Figure 7.2(b). If *portscan* alert occurs before alert *imap buffer overflow* and they have the same target IP addresses, then their pre- and post-conditions are matched. The corresponding state of leaf node *Probe/Surveillance* in Figure 7.2(b) will be set as "matched". The Bayesian-model computes the evidence and outputs the probability or confidence of the correlation of these two alerts.

3.3   Parameters in Bayesian Model

When using a Bayesian model for inference, we need to set two types of parameters, i.e., prior probability of root's states and CPT associated with each child node. In this section, we describe how we set the parameters used in our Bayesian model.

3.3.1   Parameters in Bayesian Model I: Prior Probability or Estimation on Attack Transition.

In this section, we describe the attack classes used in our work and the prior probability estimation on attack transition, i.e., the root states in our model.

The prior probability of root states (e.g., $P(correlation = high)$) used in the inference engine is set based on the attack class of alerts being correlated. It indicates the *prior knowledge* estimation of the possibility that one attack class reasonably transits to another one. For example, it is reasonable for us to have a higher estimation of the possibility that an exploit attack follows a probe than the other way around. We use domain-specific knowledge based on prior experience and empirical studies to estimate appropriate probability values. Related work [52] also helps us on the probability estimation.

In our work, we denote the attack classes as the follows.

C1: Super Privilege Violation. C2: User Privilege Violation. C3: DoS. C4: Probe. C5: Access Violation. C6: Integrity Violation. C7: Asset Distress. C8: Connection Violation. C9: Malicious Binary Installation. C10: Exfiltration.

In alert correlation, the pair of alerts being evaluated in the correlation engine (as shown in Figure 7.2(b)) is only known at run-time. Therefore, we cannot use an inference engine with a fixed set of CPT parameters. Instead, we set up a set of CPTs based on each pair of attack classes (e.g., *Malicious Concealment* and *DoS*). At run-time, when correlating a pair of alerts A_i and A_j with respective corresponding attack classes $C(A_i)$ and $C(A_j)$ (e.g., alert *imap buffer overflow* with attack class *Super PrivilegeViolation* and alert *illegal file access* with attack class *Access Violation*), the inference engine selects the corresponding CPT parameters for the attack classes $C(A_i)$ and $C(A_j)$, and computes the overall probability that A_j is "caused" by A_i given the evidence from the evaluators, i.e., $P(correlation = high|e = evidence)$. An implicit temporal constraint between these two alerts is that alert A_i occurs before A_j. In this example, we can interpret the correlation as: the *imap buffer overflow* attack is followed by an illegal access to a file after the attacker gets root privileges on the target. Initial values of CPTs are pre-defined based on our experience and domain knowledge.

We have also defined an attack transition table that includes the estimated possibility that how reasonably an attack with class C_i (i.e., i^{th} column in the matrix) may progress to another attack with class C_j (i.e., j^{th} row in the matrix). The table entry is used as the prior probability of the root state in our model.

The estimation is based on our prior experience and empirical studies. In the process of estimation, we also refer to the related work in [52].

3.3.2 Parameters in Bayesian Model II: Adaptive CPT Update.

Another important parameter in Bayesian model is the CPT associated with each node. CPT values associated with each node adapt to new evidence and therefore can be updated accordingly. We apply an adaptive algorithm originally proposed by [1] and further developed by [9]. The motivation of using adaptive Bayesian network is that we want to fine-tune the parameters of the model and adapt the model to the evidence to fix the initial CPTs that may be pre-defined inappropriately. The intuition of the algorithms proposed by [1] is that we want to adapt the new model by updating CPT parameters to fit the new data cases while balancing movement away from the current model.

Specifically, we denote X as a node in a Bayesian network, and let U be the parent node of X. X has r states with values of x_k, where $k = 1, ..., r$ and U has q states with values of u_j, where $j = 1, ..., q$. An entry of CPT of the node X can be denoted as: $\theta_{jk} = P(X = x_k | U = u_j)$. Given a set of new data cases, denoted as D, $D = y_1, ..., y_n$, and assuming there is no missing data in evidence vector of y_t, where evidence vector y_t represents the evidence at the tth time, the CPT updating rules are:

$$
\begin{aligned}
\theta_{jk}^t &= \eta + (1 - \eta)\theta_{jk}^{t-1}, \ \ for\ P(u_j|y_t) = 1\ and\ P(x_k|y_t) = 1. \quad (7.1)\\
\theta_{jk}^t &= (1 - \eta)\theta_{jk}^{t-1}, \ \ for\ P(u_j|y_t) = 1\ and\ P(x_k|y_t) = 0. \quad (7.2)\\
\theta_{jk}^t &= \theta_{jk}^{t-1}, \ otherwise. \quad (7.3)
\end{aligned}
$$

η is the learning rate. The intuition of the above updating rules is that, for an entry of CPT, e.g., θ_{mn}, we either increase or decrease its value (i.e., $P(X = x_n | U = u_m)$) based on the new evidence received. Specifically, given the evidence vector y_t, if the parent node U is observed in its mth state, i.e., $U = u_m$, and X is in its nth state, i.e., $X = x_n$, we regard the evidence as *supporting evidence* of the CPT entry θ_{mn}. We then increase its value (i.e., $P(X = x_n | U = u_m)$), which indicates the likelihood that X is in its nth state given the condition that parent node U is in its mth state, as shown in Eq. (7.1). By contrast, if node X is not in its nth state while its parent node U is in the mth state, we then regard the evidence as *un-supporting evidence* of θ_{mn} and decrease θ_{mn}'s value as shown in Eq. (7.2). We do not change the value of θ_{mn} if no corresponding evidence is received. The learning rate η controls the rate of convergence of θ. η equaling 1 yields the fastest convergence, but also yields a larger variance. When η is smaller, the convergence is slower but eventually yields a solution to the true CPT parameter [9]. We build our inference model based on above updating rules.

We also need to point out that the adaptive capability of the inference model does not mean that we can ignore the accuracy of initial CPT values. If the initial values are set with a large variance to an appropriate value, it will take time for the model to converge the CPT values to the appropriate points. Therefore, this mechanism works for fine-tuning instead of changing CPT values dramatically.

For an alert pair, (A_i, A_j), if its correlation value computed by the Bayesian-based model, denoted as P_{bayes}, is larger than a pre-defined threshold, e.g., 0.5, then we say Bayesian-based correlation engine identifies that alert A_j is "caused" by alert A_i.

3.4   Summary

Our alert correlation engine using Bayesian network has several advantages. First, we can incorporate prior knowledge and expertise by populating the CPTs. It is also convenient to introduce partial evidence and find the probability of unobserved variables. Second, it is capable of adapting to new evidence and knowledge by belief updates through network propagation. Third, the correlation output is probability rather than a binary result from a logical combination. We can adjust the correlation engine to have the maximum detection rate or a minimum false positive rate by simply adjusting the probability threshold. By contrast, it is not directly doable when using a logical combination of pre-/post-condition matching. Finally, Bayesian networks have been studied extensively and successfully applied to many applications such as causal reasoning, diagnosis analysis, event correlation in NMS, and anomaly detection in IDS. We have confidence that it can be very useful to INFOSEC alert correlation.

There are also several limitations in our approach. First, our correlation engine relies on the underlying security sensors (e.g., IDSs) to provide alerts. If the security sensors miss a critical attack that links two stages of a series of attacks, the related attack steps may be split into two correlated groups. Therefore, we need some other techniques (e.g., attack plan recognition) to link isolated alert sets that includes correlated alerts. Second, our approach is based on domain knowledge of attack transition patterns. If there are new attack transition patterns or two related alerts have no direct causal relationship, our approach is not fully effective. Therefore, we need to develop complementary correlation techniques (e.g., statistical-based correlation technique) and use them along with our Bayesian-based correlation engine.

4.   Statistical-Based Alert Correlation

4.1   Motivation

The motivation to develop another complementary correlation mechanism is to discover more attack step dependency that the prior correlation engines have missed. Our Bayesian-based correlation engine focuses on discovering alert pairs with direct causal relationship (i.e., the consequences of an earlier attack consequence satisfy or partially satisfy the prerequisite of a later attack). In order to discover attack steps that have indirect dependency but strong statistical and temporal patterns, we have developed two correlation engines based on statistical and temporal analysis. In this section, we introduce our GCT-based correlation engine using Granger Causality Test (GCT) [20]. In Section 5, we introduce our correlation mechanism based on causal discovery theory.

4.2   Time Series Analysis

Time series analysis aims to identify the nature of a phenomenon represented by a sequence of observations. The objective requires the study of patterns of the observed time series data.

There are two main goals of time series analysis: (a) identifying the nature of the phenomenon represented by the sequence of observations, and (b) forecasting (predicting future values of the time series variable). Both goals require that the pattern of observed time series data is identified and more or less formally described. Once the pattern is established, we can interpret and integrate it with other techniques to extrapolate future events.

A time series is an ordered finite set of numerical values of a variable of interest along the time axis. It is assumed that the time interval between consecutively recorded values is constant. We denote a univariate time series as $x(k)$, where $k = 0, 1, \ldots, N - 1$, and N denotes the number of elements in $x(k)$.

Time series causal analysis deals with analyzing the correlation between time series variables and discovering the causal relationships. Causal analysis in time series has been widely studied and used in many applications, e.g., economy forecasting and stock market analysis.

Granger Causality Test (GCT) is a time series-based *statistical* analysis method that aims to test if a time series variable X correlates with another time series variable Y by performing a *statistical hypothesis test*. In time series analysis theory, although there exist some other simple lagged correlation analysis, e.g., computing correlation coefficients between two time series variables, GCT has been proved to be more rigorous. GCT was originally proposed and applied in econometrics, it has been widely applied in other areas, such as weather analysis (e.g., [32]), automatic control system (e.g., [5, 18]) and neurobiology (e.g., [31, 26]).

Network security is another application in which time series analysis can be very useful. In our prior work [3, 2], we have used time series-based causality analysis for pro-active detection of Distributed-Denial-of-Service (DDoS) attacks using MIB II [51] variables. We based our approach on the Granger Causality Test (GCT) [20]. Our results showed that the GCT is able to detect the "precursor" events, e.g., the communication between Master and Slave hosts, without prior knowledge of such communication signatures, on the attacker's network before the victim is completely overwhelmed (e.g., shutdown) at the final stage of DDoS.

In this work, we apply the GCT to INFOSEC alert streams for alert correlation and scenario analysis. The intuition is that attack steps that do not have well-known patterns or obvious relationships may nonetheless have some temporal correlations in the alert data. For example, there are one or more alerts for one attack only when there are also one or more alerts for another attack within a certain time window. We can apply temporal causality analysis to find such alerts to identify an attack scenario. We next give some background on the GCT.

4.3   Granger Causality and Granger Causality Test

The intuition of Granger Causality is that if an event X is the cause of another event Y, then the event X should precede the event Y. Formally, the Granger Causality Test (GCT) uses statistical functions to test if *lagged* information on a time-series variable x provides any statistically significant information about another time-series variable y. If the answer is yes, we say variable x Granger-causes y. We model variable y by two auto-regression models, namely, the Autoregressive Model (AR Model) and the Autoregressive Moving Average Model (ARMA Model). The GCT compares the residuals of the AR Model with the residuals of the ARMA Model. Specifically, for two time series variables y and x with size N, the Autoregressive Model of y is defined as:

$$y(k) = \sum_{i=1}^{p} \theta_i y(k-i) + e_0(k) \tag{7.4}$$

The Autoregressive Moving Average Model of y is defined as:

$$y(k) = \sum_{i=1}^{p} \alpha_i y(k-i) + \sum_{i=1}^{p} \beta_i x(k-i) + e_1(k) \tag{7.5}$$

Here, p is a particular lag length, and parameters α_i, β_i and θ_i ($1 \leq i \leq p$) are computed in the process of solving the Ordinary Least Square (OLS) problem (which is to find the parameters of a regression model in order to have the minimum estimation error). The residuals of the AR Model is $R_0 = \sum_{k=1}^{T} e_0^2(k)$, and the residuals of the ARMA Model is $R_1 = \sum_{k=1}^{T} e_1^2(k)$. Here, $T = N - p$.

The AR Model, i.e., Eq.(7.4), represents that the current value of variable y is predicted by its past p values. The residuals R_0 indicate the total sum of squares of error. The ARMA Model, i.e., Eq.(7.5), shows that the current value of variable y is predicted by the past p values of both variable y and variable x. The residuals R_1 represents the sum of squares of prediction error.

The Null Hypothesis H_0 of GCT is $H_0 : \beta_i = 0, i = 1, 2, \cdots, p$. That is, x does not affect y up to a delay of p time units. We denote g as the Granger Causality Index (GCI):

$$g = \frac{(R_0 - R_1)/p}{R_1/(T - 2p - 1)} \sim F(p, T - 2p - 1) \tag{7.6}$$

Here, $F(a, b)$ is Fisher's F distribution with parameters a and b [23]. F-test is conducted to verify the validity of the Null Hypothesis. If the value of g is larger than a critical value in the F-test, then we reject the Null Hypothesis and conclude that x Granger-causes y. Critical values of F-test depends on the degree of freedoms and significance value. The critical values can be looked up in a mathematic table [24].

The intuition of GCI (g) is that it indicates how better variable y can be predicted using histories of both variable x and y than using the history of y alone. In the ideal condition, the ARMA model precisely predicts variable y with residuals $R_1 = 0$, and the GCI value g is infinite. Therefore, the value of GCI (g) represents the strength of the causal relationship. We say that variable $\{x_1(k)\}$ is more likely to be causally related with $\{y(k)\}$ than $\{x_2(k)\}$ if $g_1 > g_2$ and both have passed the F-test, where g_i, $i = 1, 2$, denotes the GCI for the input-output pair (x_i, y).

4.4   Procedure of Data Processing in GCT

Before applying GCT to data sets, we propose a procedure of data processing. In each step, there are multiple possible testing techniques and we chose the one that is most commonly used and conveniently implemented.

Step 1: testing for individual stationary. This step is to statistically test if each data set is stationary. A stationary time series means the probability distribution is stable during the stochastic process. In this step, we use testing technique proposed by Dickey-Fuller [15].

Step 2: data transformations. For non-stationary data sets, we can apply transform functions to change a non-stationary time series into a stationary one. The most common used transformations are log transformation and the differencing transformation. They can be also used together. For example, an initial log transformation is followed by first differencing, i.e., $(1 - L)Log(x(t)) = Log(x(t)) - Log(x(t-1))$, where L represents *lag operator* defined as $Lx(t) = x(t-1)$ and $(1 - L)x(t) = x(t) - x(t-1)$.

Step 3: testing for multivariate independence. This step is to test if two time series variables are *statistically independent* of each other. The available test techniques are proposed by Chitturi [8] and Hosking [27].

In practice, we can go through this step and then conduct the GCT for the non-independent bivariates. Results of GCT can tell us if they are causally related and the causal order or direction. As an alternative, we can also skip this step and conduct GCT directly because we can also infer the variable relationship from GCT output that can tell if they are independent of each other or if there are any causal relationships.

Step 4: testing for co-integration of data sets. In this step, we can apply multivariate version of Dickey-Fuller Test or Johansen Test [29] to test the existence of co-integration between two time series. Theoretically, GCT can be conducted on two co-integrated time series variables. However, as Lee et al. [35] empirically pointed out, GCT can result in spurious causality when testing co-integrated variables. Therefore, we recommend not to apply GCT on co-integrated time series in order to avoid the inaccuracy.

Step 5: testing Granger Causality. As described in Section 4.3, we conduct the statistical hypothesis test with a significance level, e.g., 5% or 1%.

Step 6: confidence computation. This step is to compute the probability or confidence of correlation. As GCI conforms to F-distribution, i.e., $F(p, T - 2p - 1)$, therefore, we can compute the corresponding probability as: $P_{gct} = CDF_{F-distribtuion}(p, T - 2p - 1, GCI)$, which represents the correlation confidence between two variables.

4.5 Applying GCT to Alert Correlation

4.5.1 Alert Time Series Formulation. Before applying GCT to alert correlation, we need to formulate each hyper alert into a univariate time series.

Specifically, we set up a series of time slots with equal time interval, denoted as t_{slot}, along the time axis. Given a time range T, we have $m = T/t_{slot}$ time slots. Recall that each hyper alert or cluster A include a set of alert instances with the same attributes except time stamps, i.e., $A = [a_1, a_2, \ldots, a_n]$, where a_i represents an aggregated alert instance in the cluster, we denote $\tilde{A}$ as the corresponding time series variable of hyper alert A. $\tilde{A} = \{n_1, n_2, \ldots, n_m\}$, where each value n_i represents the number of alert instances of hyper alert A occurring within a specific time slot $slot_i$.

Table 7.2 is an example that shows how we formulate a time series variable for each hyper alert. From Table 7.2, we can see that the time variable $\tilde{A}$'s value equals the number of alert instances of hyper alert A occurring within a time slot.

We currently do not use categorical variables such as port accessed and pattern of TCP flags as time series variables in our approach.

Table 7.2. An example of alert time series formulation

Time slot	Number of A's alert instances	$\tilde{A}$'s value
$slot_1$	1	1
$slot_2$	5	5
...	...	...
$slot_i$	9	9
$slot_m$	0	0

4.5.2 GCT-based Alert Correlation. Applying the GCT to alert correlation, the task is to determine which hyper alerts among $A_1, A_2, ..., A_l$ most likely have the causal relationship with hyper alert B (a hyper alert represents a sequence of alerts in the same cluster). Based on alert priority value and mission goals as described in Section 2.2, the security analyst can specify a hyper alert as a target (e.g., alert *Mstream_DDOS* against a database server) which other alerts are correlated with. The GCT algorithm is applied to the corresponding alert time series. The formulation of alert time series is described in Section 4.5.1.

As described in Section 4.5.1, values of a hyper alert's time series (e.g., $\tilde{B}$) represent the number of alert instances occurring within a certain time period. Specifically, given a hyper alert B, for each hyper alert pair, i.e., $(A_i, B), i = 1, 2, \ldots, m$, we apply GCT to their corresponding time series variables, i.e., $GCT(\tilde{A}_i, \tilde{B})$. In other words, we are testing the temporal correlation of alert instances to determine if $\tilde{A}_i$ has a causal relationship with $\tilde{B}$.

As described in Section 4.3, the GCT index (GCI) g returned by the GCT function represents the evidence strength of the cause-effect relationship, and GCI also conforms to F-distribution. In practice, after performing GCT computation on each pair of alert time series variables (e.g., $GCT(\tilde{A}_i, \tilde{B})$, $i = 1, 2, \ldots, m$), we record the alert time series variables whose GCI values have passed the F-distribution test (e.g., $\tilde{A}_1, \tilde{A}_5, \tilde{A}_9$), then select the corresponding hyper alerts (e.g., A_1, A_5, A_9) as candidates of causal alerts w.r.t. alert B. We rank order the candidate alerts according to their GCI values, then select the top m candidate alerts and regard them as being causally related to alert B. These candidate relationships can be further inspected by other techniques or security analyst based on expertise and domain knowledge. The corresponding attack scenario is constructed based on the correlation results.

In alert correlation, identifying and removing background alerts is an important step. We use *Ljung-Box* [37] test to identify the background alerts. The assumption is that background alerts have characteristic of randomness. The *Ljung-Box* algorithm tests for such randomness via autocorrelation plots. The Null Hypothesis is that the data is random. The test value is compared with critical values to determine if we reject or accept the Null Hypothesis.

When applying GCT, one important parameter is the variable p as shown in Eq.(7.4) and Eq.(7.5). This parameter represents the number of history values (or the length of lagged time window) needed when performing the GCT.

Given two hyper alerts A and B that have corresponding time series $\tilde{A} = \{\tilde{a}_1, \tilde{a}_2, ..., \tilde{a}_i, ..., \tilde{a}_n\}$ and $\tilde{B} = \{\tilde{b}_1, \tilde{b}_2, ..., \tilde{b}_j, ..., \tilde{b}_n\}$ respectively, we want to identify if A Granger-causes B or not. As described in Section 4.2, a time series variable is under the assumption that the time interval between consecutively recorded values is constant. Therefore, the position difference between time series instances can be regarded as the time delay between alert instances.

In our work, We denote the corresponding parameter p as $p_{\tilde{A}\tilde{B}}$. We set the parameter $p_{\tilde{A}\tilde{B}}$ as follows.

DEFINITION 2 *Given a time series variable instance $\tilde{a}_i$ ($\tilde{a}_i \in \tilde{A}$ and $\tilde{a}_i \neq 0$) and its most adjacent time series instance $\tilde{b}_j$ ($\tilde{b}_j \in \tilde{B}$, $\tilde{b}_j \neq 0$ and $j > i$), we denote $\Delta d_{i,j}$ as the adjacent time delay between $\tilde{a}_i$ and $\tilde{b}_j$.*
$$\Delta d_{i,j} = j - i$$
We denote $d_{adjacent\_time\_gap}$ as a set variable that unions all the time delays between adjacent time series instances in $\tilde{A}$ and $\tilde{B}$.
$$d_{adjacent\_time\_gap} = \bigcup \{\Delta d_{i,j}\}$$
where $i, j = 1, 2, ..., n$.

We then set $p_{\tilde{A}\tilde{B}}$ as $p_{\tilde{A}\tilde{B}} = \max\{d_{adjacent\_time\_gap}\}$.

The intuition of the method of setting parameter p is that we want to have a time window with an enough length so that we can include all potential causal alerts with respect to an effect alert.

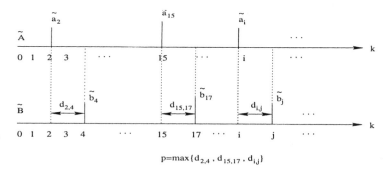

$$p=\max\{d_{2,4}, d_{15,17}, d_{i,j}\}$$

Figure 7.3.   An example of time delay between time series instances

Figure 7.3 shows an example how we set the parameter p. In the figure, time series variable $\tilde{A}$ has 3 non-zero instances at $k = 2, 15, i$ (i.e., $\tilde{a}_2, \tilde{a}_{15}, \tilde{a}_i$), time series variable $\tilde{B}$ has 3 non-zero instances at $k = 4, 17, j$ (i.e., $\tilde{b}_4, \tilde{b}_{17}, \tilde{b}_j$). We set p as the maximum value of delays between adjacent time series instance as shown in Figure 7.3.

The main advantage of using statistical causality test such as GCT for alert correlation is that the approach does not require *a priori* knowledge about attack behaviors and how the attacks could be related. This approach can identify the correlation between two attack steps as long as the two have a temporal pattern (not necessarily high frequency) when occurring together. We believe that a large number of attacks, e.g., worms, have attack steps with such characteristics. Thus, we believe that causal analysis is a very useful technique. As discussed in [3, 2, 4], when there are sufficient training data available, we can use GCT off-line to compute and validate very accurate causal relationships from alert data. We can then update the knowledge base with these "known" correlations for efficient pattern matching in run-time. When GCT is used in real-time and finds a new causal relationship, as discussed above, the top m candidates can be selected for further analysis by other techniques.

5. Causal Discovery-Based Alert Correlation

5.1 Motivation

Knowledge-based alert correlation system depends on attack transition patterns to correlate security alerts. It has the advantage of efficiency and accuracy. However, the signature-based correlation system lacks the capability of detecting the attack transitions whose scenario patterns are unknown. In practice, security analysts are more interested in those *novel* attack strategies that can easily evade signature-based correlation analysis and can potentially cause more damages due to the lack of knowledge about them.

Bearing this challenge in mind, we have studied and built a correlation technique based on statistical analysis. This correlation engine is based on the hypothesis that for some attack steps, even though they do not have direct causal relationship, they can have statistical dependence patterns. For example, a malicious daemon keeps uploading sensitive information to an external site and downloading new malicious code updates from the external site. GCT-based correlation engine as described in Section 4 is an approach to identify this type of attack transition patterns. GCT-based correlation technique has the strength of identifying the causality direction between alert pairs, however, it has the limitation on attack steps that have a weak temporal pattern (e.g., the variance of time lags between attack steps is large). In order to identify the statistical dependence pattern of alert pairs that have weak temporal relationship, we have developed a correlation mechanism based on causal discovery theory [44]. Our goal is to identify *new* attack transition patterns beyond the limitation of domain-knowledge.

In this section, we introduce and describe our correlation mechanism using causal discovery theory.

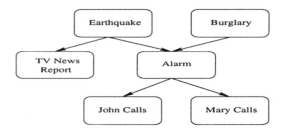

Figure 7.4. An example of a causal network

5.2   Introduction to Causal Discovery

Causal discovery has been an active research topic in the fields of artificial intelligence (AI) and social science. The goal of causal discovery is to test and identify causal relationships among variables under study. Researchers have developed and shown that causal Bayesian network can be used to represent the causal relationships between variables [44].

5.2.1   Causal Bayesian Network.   A Bayesian network is usually represented as a directed acyclic graph (DAG) where each node represents a variable, and the directed edges represent the causal or dependent relationships among the variables.

Figure 7.4 shows an example of a causal network adapted from [43]. Here, a house alarm may sound as a result of either a burglary or an earthquake. An earthquake may also result in a TV news report. Neighbors John or Mary may report a call when the alarm sounds. The directed edge represents the cause-effect between variables.

In practice, causal discovery can be regarded as a task of constructing causal Bayesian networks from observational data.

Learning a Bayesian network from data includes two subtasks, i.e., learning the structure of the Bayesian network and learning the parameters of the network. The first subtask learns the causal relationship between variables and the second one represents the strength of these dependencies, which are encoded in conditional probability tables (CPTs) associated with each child node. Specifically, an element of the CPT at a child node is a conditional probability defined as $CPT_{ij} = P(child\_state = j | parent\_state = i)$ [43]. Since it is relatively straightforward to learn the parameters given observational data and a causal network structure, the challenge in causal discovery is the first task, i.e., learning the network structure from data sets.

In the causal discovery theory, the fundamental assumption is *causal Markov condition*. Causal Markov condition means that, in a causal Bayesian network, any node is conditionally independent of its non-descendants (i.e., non-effect nodes) given its parent nodes (i.e., direct causes) [50]. The independence relationships represented by the structure of a causal Bayesian network are given by the causal Markov condition.

The conditional independence properties of a causal network can be deduced from the structure of the DAG by the *d-separation* criterion as defined in [43].

The structure of a causal network under discovery is a directed acyclic graph (DAG) that encodes conditional independencies via the causal Markov assumption. Learning the Bayesian network structure from the data actually is the process of identifying the conditional independency among variables.

5.2.2   Approaches to Causal Discovery.   Based on causal Markov assumption, there have been many research work on causal discovery. Generally, there are two approaches to discovering causal Bayesian networks.

One causal discovery approach is based on score functions, e.g., Bayesian computation [10, 16, 25]. Intuitively, this approach computes the probability that the causal relationship exists among the variables. For each pair of variables, a probabilistic computation is conducted to exam the dependence or independence between the two variables. In looking for the structures that fit for the conditional independence constraints, the approach in [25] makes *probabilistic* inferences about the conditional-independence constraints and the goal is to find the Bayesian network structures that have maximum score.

In [25], the score is defined as the posterior probabilities $p(m|D)$, where m corresponds to the causal network models learned from the given data D. This Bayesian-based approach can give a quantitative evaluation of causal network structures constructed from data. The goal is to identify a causal network structure $\tilde{m}$ ($\tilde{m} \in m$) so that $p(\tilde{m}|D)$ has the maximum value among all other causal network structures learned from data D. One challenge to this approach is model search and selection. Researchers usually use *model selection* method to select the best fitted model among others (i.e., the one with highest posterior probability $p(m|D)$) or *selective model averaging* method to average a number of better fitted models from all models [25]. There are still challenges to these two model selection methods, in particular, the accuracy issue [25]. In practice, people use some heuristic search algorithms to solve the model selection problems. However, those heuristic search algorithms may not give the best causal Bayesian network structures. Some scoring-based algorithms also have the issues that the different input ordering of variables can generate very different causal network structures.

Another category of causal discovery mechanism is *constraint-based* or *dependency analysis-based* approach (e.g., [6, 50]). This category of approaches usually apply statistical tests (e.g., χ^2 test, a statistical test for accepting or rejecting an hypothesis [24], and mutual information, a measure of dependency between variables [11]) to discovering conditional independence and dependence among variables and use these relationships as constraints to construct a Bayesian network. Specifically, for each pair of variables, this approach tests if any dependence exists. If so, an edge will be added between these two variables accordingly. Further tests will be conducted on each edge to examine if the two end-nodes are found to be conditionally independent. If the conditional independence is identified, the edge will be removed. The intuition on this approach is that a pair of nodes with larger test score (e.g., mutual information that measures the dependency between variables) is more likely to represent a direct connection (an edge) than a pair with smaller test score, which may represent an indirect connection. Search and scoring methods can be applied to identifying the directions of edges.

In our work, we applied *constraint-based* approach using mutual information for conditional independence test [6].

In information theory [11], *mutual information* is defined and used to measure the statistical dependence between two random variables.

DEFINITION 3 *For two random variables X and Y with a joint probability distribution $P(x, y)$ and marginal probability distributions $P(x)$ and $P(y)$, mutual information $I(X, Y)$ is defined as [11]*

$$I(X, Y) = \sum_{x,y} P(x, y) \log \frac{P(x, y)}{P(x)P(y)} \tag{7.7}$$

DEFINITION 4 *For three random variables X, Y and Z with a joint probability distribution $P(x, y, z)$ and conditional probability distributions $P(x, y|z)$, $P(x|z)$ and $P(y|z)$, the conditional mutual information $I(X, Y|Z)$ is defined as [11]*

$$I(X, Y|Z) = \sum_{x,y,z} P(x, y, z) \log \frac{P(x, y|z)}{P(x|z)P(y|z)} \tag{7.8}$$

Intuitively, mutual information $I(X, Y)$ measures the information of X that is shared by Y. If X and Y are independent, then X contains no information about Y and vice versa, so their mutual information is zero. If X and Y are dependent, knowing the value of one variable can give us some information about the value of the other. In building the causal Bayesian network, we can apply mutual information to test if two variables are dependent and evaluate the strength of corresponding dependence.

Similarly, *conditional mutual information* $I(X, Y|Z)$ is used to test if two variables (i.e., X and Y) are dependent given the condition variable Z.

In theory, we claim X and Y are independent when $I(X, Y) = 0$ given the actual distributions of corresponding variables. In practice, given a data set D, we use empirical instead of theoretic distributions of variables when computing mutual information. Therefore, the normal practice is usually to set up a small threshold ϵ and claim X and Y are independent when $I(A, B) < \epsilon$.

Similarly, we declare X and Y are conditionally independent given Z when $I(X, Y|Z) < \epsilon$.

The intuition of applying mutual information to causal discovery is that when two variables have a strong statistical dependency pattern, mutual information can detect it. The causality direction is determined by the conditional mutual information measure under the assumption that, in a causal network, two cause nodes are independent with each other, but conditionally dependent with each other given a common effect node. Specifically, based on mutual information measure, if we have identified variable A, B are mutually independent (i.e., $I(A, B) < \epsilon$), and are dependent with C respectively (i.e., $I(A, C) > \epsilon$, $I(B, C) > \epsilon$), and if we have also identified variable A and B are conditionally dependent given C based on conditional mutual information measure (i.e., $I(A, B|C) > \epsilon$), then we can determine that A, B are causes to C, i.e., $\{A \rightarrow C, B \rightarrow C\}$. Such structure is called **V-structure** [25]. Such determination intuitively satisfies the notion of causality because when an effect is determined (i.e., given C), the increasing confidence on cause A reduces the belief that B causes C. In our example of Figure 7.4, we have seen such cause-effect pattern among *Earthquake, Burglary* and *Alarm*.

In our work, we did not to select the score function-based approach (e.g., [25]) because that approach usually requires prior knowledge (e.g., prior probability) of causal network models in the model construction, model comparison and final model selection. According to our experience it is difficult to get such prior knowledge in the security application. In fact, our goal is to identify novel attack transition patterns that can be totally unknown in the past. In [6], the researchers have developed algorithms to avoid complex conditional independence tests based on mutual information divergence. The enhanced test algorithms have eliminated the need for an exponential number of conditional independence tests that is an issue in earlier constraint-based algorithms.

5.3 Applying Causal Discovery Analysis to Alert Correlation

Before we apply causal-discovery approach to alert correlation, raw alerts need to get aggregated and clustered into *hyper alerts* as described in Section 2.1 so that we can investigate the statistical patterns between alerts.

After the above process, we formulate transaction data for each hyper alert. Specifically, we set up a series of time slots with equal time interval, denoted as t_{slot}, along the time axis. Given a time range T, we have $m = T/t_{slot}$ time slots. Recall that each hyper alert A includes a set of alert instances with the same attributes except time stamps, i.e., $A = [a_1, a_2, \ldots, a_n]$, where a_i represents an alert instance in the cluster. We denote $N_A = \{n_1, n_2, \ldots, n_m\}$ as the variable to represent the occurrence of hyper alert A during the time range T, where n_i is corresponding to the occurrence (i.e., $n_i = 1$) or un-occurrence (i.e., $n_i = 0$) of the alert A in a specific time slot $slot_i$. In other words, if there is one or more instances of alert A (e.g., a) occurring in the time slot $slot_i$, then $n_i = 1$; otherwise, $n_i = 0$.

Using the above process, we can create a set of transaction data and input them to the causal discovery engine for analysis. Table 7.3 shows an example of the transaction data corresponding to hyper alert A, B and C. The correlation engine will output the causal network model based on transaction data set.

Table 7.3. An example of transaction data set

Time slot	$Alert_A$	$Alert_B$	$Alert_C$
$slot_1$	1	0	1
$slot_2$	0	0	1
...	...	...	...
$slot_i$	1	0	0
$slot_m$	0	0	1

Algorithm 1 shows the steps to apply causal discovery theory to correlating alerts. In step 1, we apply mutual information measure to identify alerts with strong statistical dependence. In step 2, we identify alert triplets that have a *V-structure* (i.e., $X \rightarrow Z, Y \rightarrow Z$, as described in Section 5.2). The causality directions in a *V-structure triplets* are determined by the conditional mutual information measure under the assumption that, in a causal network, two cause nodes are respectively dependent with a common effect node. These two cause nodes are mutually independent with each other, but conditionally dependent with each other given a common effect node. In step 3, for the partially directed alert triplets, since A_m and A_k are not directly connected, it means A_m and A_k are mutually independent (otherwise they should have been connected in step 1). The causality direction between A_n and A_k is tested based on the causal Markov assumption (i.e., in a causal network, a node X is independent to other nodes (except its direct effect node) given X's direct cause). Therefore, if A_m and A_k are also conditionally independent given A_n, we can identify the causality direction between A_n and A_k (i.e., $A_n \rightarrow A_k$). Otherwise, if A_m and A_k are conditionally dependent given A_n, the triplet has a *v-structure*, then A_k is the parent node of A_n (i.e., $A_k \rightarrow A_n$).

Figure 7.5 shows an example of the causal network model among alert A, B and C of which A and B are two causal alerts of C. As described in Section 5.2.1, in a causal network, each non-root node is associated with a conditional probability table (CPT) that shows the strength of the causal relationship between the node and its parent node. Table 7.4 shows the CPT entries associated with alert C in which "1" represents the occurrence of the alert and "0" represents the nonoccurrence. Among the CPT entries as shown in Table 7.4,

---

Algorithm 1 Alert correlation using causal discovery theory

---

1. For each alert pair A_i, A_j

if A_i and A_j are dependent using mutual information measure, i.e., $I(A_i, A_j) > \epsilon$, where ϵ is a small threshold, **then**

    Connect A_i and A_j directly.

end if

2. For any three alerts A_m, A_n, A_k that have the connection pattern that A_m and A_n, A_n and A_k are directly connected, and A_m and A_k are not directly connected (i.e., $A_m - A_n - A_k$)

if A_m and A_k are conditionally dependent given A_n using conditional mutual information measure, i.e., $I(A_m, A_k|A_n) > \epsilon$ **then**

    Let A_m be the parent node of A_n, and A_k be the parent node of A_n, respectively, (i.e., $A_m \rightarrow A_n$ and $A_k \rightarrow A_n$).

end if

3. For any three alerts A_m, A_n, A_k that have a partially directed pattern $(A_m \rightarrow A_n - A_k)$, i.e., A_m is a parent node of A_n, A_n and A_k are directly connected (edge (A_n, A_k) is not oriented), and A_m is not directly connected with A_k

if A_m and A_k are conditionally independent given A_n, i.e., $I(A_m, A_k|A_n) < \epsilon$, **then**

    Let A_n be the parent node of A_k, i.e., $A_n \rightarrow A_k$.

else if A_m and A_k are conditionally dependent given A_n, i.e., $I(A_m, A_k|A_n) > \epsilon$, **then**

    Let A_k be the parent node of A_n, i.e., $A_k \rightarrow A_n$.

end if

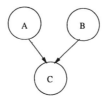

Figure 7.5. An example of the causal network model of alert A, B and C

Table 7.4. An example of CPT associated with node C

	$AB = 0$	$AB = 01$	$AB = 10$	$AB = 11$
$C = 0$	p_1	p_2	p_3	p_4
$C = 1$	p_5	p_6	p_7	p_8

we are more interested in p_6 and p_7. The value of p_6 represent the probability of the occurrence of alert C given that alert B has already occurred, i.e., $p_6 = P(C = 1|B = 1)$. Similarly, the entry of p_7 shows the dependency of alert C to the causal alert A, i.e., $p_7 = P(C = 1|A = 1)$. In practice, we can regard p_6 and p_7 as the likelihood of attack step transition from attack B to attack C and from attack A to attack C, respectively.

Given the transaction data, computing the CPT entries is more straightforward. For example, the value of p_6 can be empirically computed as $P(C = 1|B = 1) = \frac{\# \ of \ (B=1,C=1)}{\# \ of \ (B=1)}$. We can also apply the algorithm of adaptive CPT updates as described in Section 3.3.2 to update the parameters.

6. Integration of Three Correlation Engines

6.1 The Integration Process of Three Correlation Engines

Our three correlation engines are built on different techniques and focus on different correlation aspects. Bayesian-based correlation engine is analogous to an extension of pattern matching-based detection. Causal discovery theory-based correlation mechanism investigates statistical pattern of attack step occurrences to identify causal relationship between alerts. GCT-based correlation engine focuses on temporal pattern of attacks to discover new attack transition patterns.

The rationale of our integration process in alert correlation is analogous to intrusion detection where security analysts usually first apply *pattern-based detection*, then *anomaly detection* to cover the attack space that pattern-matching method cannot discover.

In practice, we integrate and apply the three correlation mechanisms with the following steps.

First, we apply Bayesian-based correlation engine on target hyper alerts. Target alerts are hyper alerts with high priorities computed by the *alert priority computation module* as described in Section 2.2. Thus, they should be the main interests in the correlation analysis to correlate with all the other hyper alerts. The goal of this step is to correlate alerts that have direct relationship based on prior knowledge of attack step transitions. The result of this step can be a set of isolated correlation graphs. For those alert pairs that have not got any causal relationship, we leave them to be processed in the next step.

Second, for those uncorrelated alert pairs, we run causal discovery-based correlation engine to correlate them. The goal of this step is to discover more correlation between alerts that have not been identified in the prior step.

Third, for each alert pair that has not established any cause-effect relationship from prior correlation engines, we apply GCT to it. That is, GCT is used to correlate alerts that have strong temporal relationship and link the isolated correlation results together.

Figure 7.6 shows an example of our integration process. For example, we have 8 hyper alerts, denoted as $A_1, A_2, A_3, A_4, A_5, A_6, A_7, A_8$. Assuming we have identified alert A_2 and A_5 as target alerts and we want to identify causal alerts w.r.t. A_2 and A_5 respectively. After applying Bayesian-based correlation engine, i.e., the first step of correlation, we have got two groups of correlated alerts, i.e., $\{A_1 \rightarrow A_2, A_3 \rightarrow A_2\}$ and $\{A_4 \rightarrow A_5\}$, as shown by solid lines in Figure 7.6. We then apply causal discovery algorithm to the rest isolated alerts that have not been correlated with A_2 and A_5 respectively. In particular, we check if causal relationship exists between alerts $\{A_1, A_5\}$, $\{A_2, A_5\}$, $\{A_3, A_5\}$, $\{A_6, A_5\}$, $\{A_7, A_5\}$, $\{A_8, A_5\}$, $\{A_4, A_2\}$, $\{A_5, A_2\}$, $\{A_6, A_2\}$, $\{A_7, A_2\}$ and $\{A_8, A_2\}$. Assuming after this step, we have got 3 more causal-related alert pairs, i.e., $\{A_3 \rightarrow A_5\}$, $\{A_6 \rightarrow A_2\}$, $\{A_4 \rightarrow A_2\}$ as represented by dotted lines in the figure. We finally apply GCT to check if the rest isolated alert pairs $\{A_1, A_5\}$, $\{A_2, A_5\}$, $\{A_6, A_5\}$, $\{A_7, A_5\}$, $\{A_8, A_5\}$, $\{A_4, A_2\}$, $\{A_5, A_2\}$ and $\{A_8, A_2\}$ have the causality w.r.t. A_5 and A_2 respectively. Figure 7.6 shows that GCT identifies the causality of $\{A_7 \rightarrow A_2\}$ and $\{A_8 \rightarrow A_5\}$ as shown by the dashed line.

6.2   Probability/Confidence Integration

In Section 3.2, we introduced our Bayesian-based correlation engine that outputs the correlation probability or confidence of two alerts, denoted as P_{bayes}. In practice, we have a threshold t, and when P_{bayes} is over the threshold t, we say the corresponding alert pair has a causal relationship identified by the Bayesian-based correlation engine.

As described in Section 5.3, the CPT associated with each child node in a causal network shows the strength of relationship between the child node and its parent node. Particularly, one CPT entry (i.e., $P(childnode = 1|parentnode = 1)$ can be interpreted as the probability of attack transition from parent node

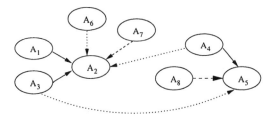

Figure 7.6. An example of integration process. The solid line represents a correlation identified by Bayesian-based correlation engine. The dotted line shows the causal relationship found by causal discovery-based correlation engine. The dashed line represents a new correlation specified by GCT-based correlation engine.

(attack) to child node (attack), e.g., the p_6 or p_7 in Table 7.4. We denote such attack transition probability as $P_{causal-discovery}$.

As discussed in Section 4.3, GCT Index (GCI) represents the strength of correlation between two alerts being correlated. It conforms to F-distribution with parameters of p and $N - 3p - 1$, where p is the number of history values of the time series variable used in the GCT computation, and N is the size of the time series variable. Therefore, for any two correlated alerts identified by GCT-based correlation engine, we can compute the corresponding F-distribution probability values, i.e., $P_{gct} = CDF_{F-distribution}(p, N - 3p - 1, GCI)$, where CDF represents the *cumulative distribution function*. P_{gct} represents the probability/confidence of correlation between two alerts.

When integrating the three correlation engines, we can adjust the confidence output from GCT-based engine as:

$$P_{gct\_final} = (P_{gct} - t) * \omega + t \qquad (7.9)$$

In Eq. (7.9), t is the threshold defined in Bayesian-based correlation engine, and ω is a weight value that is determined based on prior experience and performance measurements of the two correlation engines. The adjusted value of P_{gct_final} is in the range of $[0, t + \epsilon]$, where ϵ is a small positive number. The intuition of this adjustment is that we want to downgrade the output of GCT-based correlation engine a little because it is based on temporal analysis that is less accurate than the domain-knowledge-based Bayesian correlation engine.

Therefore, for a correlated alert pair, e.g., (A_i, A_j), we can have a probability or confidence of its correlation (i.e., attack transition from A_i to A_j) computed by Bayesian correlation engine (i.e., P_{bayes}), causal discovery algorithm (i.e., $P_{causal\_discovery}$) or GCT-based correlation mechanism (i.e., P_{gct_final}) depending on which correlation engine identifies the causal relationship. We denote the probability of alert correlation (or attack transition) as $P_{correlation}(A_i, A_j)$, i.e.,

$$P_{correlation}(A_i, A_j) = \begin{cases} P_{bayes}, & \text{if causality found by Bayesian} \\ & \text{-based correlation engine} \\ P_{causal\_discovery}, & \text{if causality found by causal} \\ & \text{discovery-based correlation} \\ & \text{engine} \\ P_{gct\_final}, & \text{if causality found by GCT} \\ & \text{based correlation engine} \end{cases}$$

We also note that two different approaches have been proposed to integrate isolated correlation graphs. Ning [40] et al. apply graph theory to measure and merge similar correlation graphs. In [41], Ning et al. link isolated correlation graphs based on attack pre-/post-conditions. Our approach is different from their work in that our integration method is based on the correlation probability evaluated by our three complementary correlation engines instead of graph or pre- /post-condition-based merging algorithms.

6.3 Attack Transition Table Updates

Statistical and temporal-based alert correlation has the advantages of discovering attack transition steps without depending on prior domain knowledge. However, compared with pattern-matching correlation techniques, it is has relatively high positive false rate and the computation cost is also relatively high.

In practice, we periodically incorporate newly discovered attack transition patterns into our domain knowledge so that we can use our Bayesian -based correlation engine to analyze and correlate alerts efficiently. Also based on new analysis results and data sets, we update the attack transition table as described in Section 3.3.1.

Denote θ as an original entry in the attack transition table, θ' as the corresponding new value computed based on new analysis results and data after a regular period T, the current table update policy is that we do not update the table entry until the new value θ' has varied from θ by a certain percentage β, e.g., 5%.

6.4 Attack Strategy Analysis

Attack strategy analysis is an important component in a correlation system. It can provides security analysts an aggregated information about what has happened and what is happening to the protected IT infrastructure.

Having correlated alert pairs output by correlation engines, we can construct attack scenarios represented by correlation graph to represent the attack strategies. A correlation graph is defined as a directed graph where each edge E_{ij} represents a causal relationship from alert A_i to A_j. Alerts with causal relationship compose the nodes in the scenario graph. We denote the node

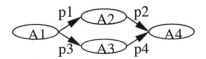

Figure 7.7. An example of correlation graph

corresponding to the causal alert as *cause node*, and the node corresponding
to the effected alert as *effect node*. A threshold t is pre-defined and alert A_j
is considered to be caused by alert A_i only when $P_{correlation}(A_i, A_j) > t$. In
constructing scenario graphs, we only include the correlated alert pairs whose
$P_{correlation}$ values are over the threshold t.

In a correlation graph, each edge is associated with a correlation probability
(i.e., $P_{correlation}$) from *cause node* to *effect node*, which can be also regarded
as the probability of attack step transition. Having such information, we can
perform *quantitative* analysis on the attack strategies. In a correlation graph,
each path is potentially a subsequence of an attack scenario and can be seen
as a Markov chain [17, 49]. Having the probability associated with each edge,
for any two nodes in the graph that are connected by multiple paths, we can
compute the overall probability of each path [49].

In the example of Figure 7.7, nodes A_1 and A_4 have to paths to connect each
other. Assuming the conditional independence of A_4 and A_1, we can compute
the overall probability of each path, e.g., $P(A_1, A_2, A_4) = P(A_4|A_2)P(A_2|A_1)$
$P(A_1) = p_2 * p_1 * p_{A_1}$.

We then rank order and select the path(s) with the highest overall correlation
probability as the most likely sequence(s) connecting two nodes.

Combining all the probability along each edge, we can also compute an
overall probability of two nodes connected with multiple paths. For example,
in the Figure 7.7, $P\{A_1 \text{ to } A_4\} = 1 - (1 - p_1 * p_2)(1 - p_3 * p_4)$.

7.   Experiments and Performance Evaluation

7.1   The Grand Challenge Problem (GCP)

To evaluate the effectiveness of our alert correlation mechanisms, we applied
our correlation algorithms to the data sets of the Grand Challenge Problem
(GCP) version 3.1 provided by DARPA's Cyber Panel program [13, 22]. In this
section, we describe and report our experiment results.

GCP version 3.1 is an attack scenario simulator. It can simulate the behavior
of security sensors and generate alert streams. GCP 3.1 includes two innovative worm attack scenarios to specifically evaluate alert correlation techniques.
In GCP, multiple heterogeneous security systems, e.g., network-based IDSs,
host-based IDSs, firewalls, and network management systems, are deployed in
several network enclaves. Therefore, GCP alerts are from both security systems and network management system. In addition to the complicated attack

scenarios, the GCP data sets also include many background alerts that make alert correlation and attack strategy detection more challenging. GCP alerts are in the Intrusion Detection Message Exchange Format (IDMEF) defined by IETF [21].

According to the GCP documents that include detailed configurations of protected networks and systems, we established a configuration database. Information on mission goals enables us to identify the servers of interest and assign interest score to corresponding alerts targeting at the important hosts. The alert priority is computed based on our model described in Section 2.2.

To better understand the effectiveness of our correlation system, we have defined two performance measures, *true positive correlation rate* and *false positive correlation rate.*

$$\begin{aligned} &True\ positive\ correlation\ rate \\ &= \frac{\natural\ of\ correctly\ correlated\ alert\ pairs}{\natural\ of\ related\ alert\ pairs} \end{aligned} \tag{7.10}$$

and

$$\begin{aligned} &False\ positive\ correlation\ rate \\ &= \frac{\natural\ of\ incorrectly\ correlated\ alert\ pairs}{\natural\ of\ correlated\ alert\ pairs} \end{aligned} \tag{7.11}$$

In Eq.(7.10), *related alert pairs* represents the alerts that have cause-effect relationship. In Eq.(7.11), *correlated alert pairs* refer to the correlation result output by a correlation system.

True positive correlation rate examines the completeness of alert correlation techniques. It measures the percentage of related alert pairs that a correlation system can identify. It is analogous to *true positive rate* or *detection rate* commonly used in intrusion detection.

False positive correlation rate measures the soundness of an alert correlation system. It examines how correctly the alerts are correlated. It is analogous to *false positive rate* used in intrusion detection.

In our experiments, we refer to the documents with the ground truth to determine the correctness of the alert correlation. Scenario graph is constructed based on alerts that have causal relationship identified by our correlation engines.

In formulating hyper alert time series, we set the unit time slot to 60 seconds. In the GCP, the entire time range is 5 days. Therefore, each hyper alert A, its corresponding time series variable $\tilde{A}$ has a size of 7,200 instances, i.e., $\tilde{A} = \{\tilde{a}_0, \tilde{a}_1, ..., \tilde{a}_{7,199}\}$.

7.1.1 GCP Scenario I. In the GCP Scenario I, there are multiple network enclaves in which attacks are conducted separately. The attack scenario in each network enclave is almost same. We select a network enclave as an example to show the correlation process.

The procedure of alert correlation is shown as follows.

First, **alert aggregation**. We conducted raw alert aggregation and clustering in order to have aggregated hyper alerts. In scenario I, there are a little more than 25,000 low-level raw alerts output by heterogeneous security devices in all enclaves. After alert fusion and clustering, we have around 2,300 hyper alerts. In our example network enclave, there are 370 hyper alerts after low-level alert aggregation.

Second, **alert noise detection**. We applied the *Ljung-Box* statistical test [37] with significance level $\alpha = 0.05$ to all hyper alerts in order to identify background alerts. In scenario I, we identified 255 hyper alerts as background alerts using this mechanism. Most of background alerts are "HTTP_Cookie" and "HTTP_Posts". Therefore, we have 115 non-noise hyper alerts for further analysis.

Third, **alert prioritization**. The next step is to select the alerts with high priority values as the target alerts. The priority computation is described in Section 2.2. In this step, we set the threshold $\beta = 0.6$. Alerts with priority scores above β were regarded as important alerts and were selected as target alerts of which we had much interest. In this step, we identified 15 hyper alerts whose priority values are above the threshold

Fourth, **alert correlation**. When applying correlation algorithms, we correlated each target alert with all other non-background alerts (i.e., the background alerts identified by the *Ljung-Box* test are excluded.). As described in Section 6.1, we have three steps in correlating alerts. First, we applied Bayesian-based correlation engine on hyper alerts and discover the correlated alert pairs. Figure 7.8 shows the correlation results related to the hyper alerts that we identified as most interested alerts. Second, we applied causal discovery-based correlation engine to alerts that have not been identified to be correlated with others in the first step. Third, we applied GCT-based correlation algorithm to further correlate alert pairs which have not been correlated after prior two steps. Figure 7.9 shows the correlation results after the three-step correlation process. The dotted line in Figure 7.8 and Figure 7.9 represent false positive correlation. The correlation probability or confidence of each alert-pair is associated with the edge in the correlation graph. In Eq. (7.9), ω equals 0.3 and t equals 0.6.

Fifth, **attack path analysis**. As discussed in Section 6.4, for any two nodes in the correlation graph that are connected on multiple paths, we can compute the probability of attack transition along each path, then rank and select the one with highest overall value. For example, from node *DB_FTP_Globbing_Attack* to node *DB_NewClient* in the graph shown in Figure 7.9, there are 6 paths that connect these two nodes. Based on the probability or confidence associated on the edge, we can compute the value of each path and rank the order.

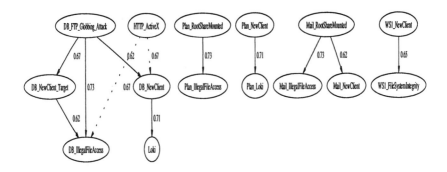

Figure 7.8. The GCP scenario I: The correlation graph discovered by Bayesian-based approach.

For example, the overall confidence for the attack path *DB_FTP_Globbing_Attack→ Loki→ DB_NewClient* is:

$P(DB\_FTP\_Globbing\_Attack, Loki, DB\_NewClient)$
$= P(DB\_FTP\_Globbing\_Attack) * P(Loki|DB\_FTP\_Globbing\_Attack)$
$*P(DB\_NewClient|Loki)$
$= P(DB\_FTP\_Globbing\_Attack) * 0.7 * 0.72$
$= P(DB\_FTP\_Globbing\_Attack) * 0.5$

Table 7.5 shows the ordered multi-paths according to the corresponding path values. From the table, we can see that it is more confident to say that the attacker is more likely to launch *FTP Globbing Attack* against the Database Server, then *New Client* attack from the Database Server that denotes a suspicious connection to an external site (e.g., set up a covert channel).

Sixth, **attack strategy analysis**. In this phase, we performed attack strategy analysis by abstracting the scenario graphs. Instead of using hyper alerts representing each node, we used the corresponding attack class (e.g., *DoS* and *Access Violation*) to abstractly present attack strategies. While analyzing attack strategy, we focused on each target and abstracted the attacks against the target. Figure 7.10(a) shows the high-level attack strategy on the Plan Server extracted from attack scenario graphs shown in Figure 7.9. From Figure 7.10(a), we can see that the attacker uses a covert channel (indicated by *Connection Violation*) to export data and import malicious code to root the Plan Server. The attacker accesses to the data stored on the Plan Server (indicated by *Access Violation*) to steal the data, then export the information. The activity of *Surveillance* has impacted the server on the performance (indicated by *Asset Distress*). Figure 7.10(b) shows the attack strategy on the Database Server. It is easy to see that the attacker launches an exploit attack against the Database Server in order to get root access. Then the attacker sets up a covert channel, accesses data and exports the data. The mutual loop pattern between attack classes *Connection Violation*, *Access Violation* and *Exfiltration* indicates the attack continuously accesses file, exports data and downloads the malicious code.

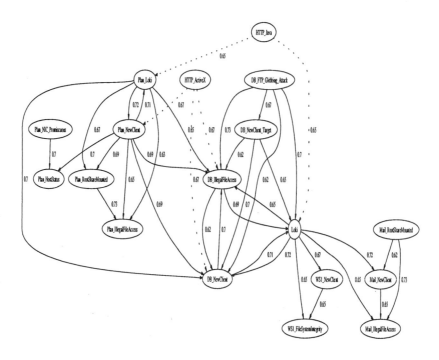

Figure 7.9. The GCP scenario I: The correlation graph discovered by the integrated approach.

Table 7.5. Ranking of paths from node *DB FTP Globbing Attack* to node *DB NewClient*.
$P = P(DB\ FTP\ Globbing\ Attack)$

Order	Nodes Along the Path	Score
Path1	DB FTP Globbing Attack→DB NewClient	P*0.62
Path2	DB FTP Globbing Attack→Loki→DB NewClient	P*0.50
Path3	DB FTP Globbing Attack→DB NewClient Target→DB NewClient	P*0.47
Path4	DB FTP Globbing Attack→DB IllegalFileAccess→DB NewClient	P*0.45
Path5	DB FTP Globbing Attack→DB NewClient Target→Loki →DB NewClient	P*0.31
Path6	DB FTP Globbing Attack→DB NewClient Target→ DB IllegalFileAccess → DB NewClient	P*0.23

7.1.2   Discussion on GCP Scenario I.

Applying our integrated correlation mechanism can discover more attack step relationships than using a single approach. Figure 7.8 shows that when we apply Bayesian-based approach alone, we can only discover partial attack step relationships. The reason is that the Bayesian-based correlation engine relies on domain knowledge to correlate alerts. Therefore, it is only capable of discovering the direct at-

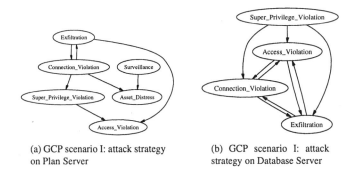

(a) GCP scenario I: attack strategy on Plan Server

(b) GCP scenario I: attack strategy on Database Server

Figure 7.10. GCP I: Attack strategy graph

tack step transitions, e.g., attack *Mail_RootShareMounted* followed by attack *Mail_IllegalFileAccess*. When the alert relationship is new or has not been encoded into the correlation engine, such relationship cannot be detected. Figure 7.9 shows that we can discover more attack relationships after applying causal discovery-based and GCT-based correlation methods. Using complementary correlation engines enable us to link isolated correlation graphs output by Bayesian-correlation engine. The reason is that our statistical and temporal-based correlation mechanisms correlate attack steps based on the analysis of statistical and temporal patterns between attack steps. For example, the loop pattern of attack transitions among attack *DB_NewClient*, *DB_IllegalFileAccess* and *Loki*. This correlation engine does not rely on prior knowledge. By incorporating the three correlation engines, in this experiment, we can improve the true positive correlation rate from 95.06% (when using GCT-based correlation engine alone [46]) to 97.53%. False positive correlation rate is decreased from 12.6% (when using GCT-based correlation engine alone [46]) to 6.89%.

Our correlation approach can also correlate non-security alerts, e.g., alerts from network management system (NMS), to detect attack strategy. Although NMS alerts cannot directly tell us what attacks are unfolding or what damages have occurred, they can provide us some useful information about the state of system and network health. So we can use them in detecting attack strategy. In this scenario, NMS outputs alert *Plan_Host_Status* indicating that the Plan Server's CPU is overloaded. Applying our GCT-based and Bayesian-based correlation algorithms, we can correlate the alert *Plan_HostStatus* with alert *Plan_NewClient* (i.e., suspicious connection) and *Plan_NIC_Promiscuous* (i.e., traffic surveillance).

7.2   GCP Scenario II

In GCP scenario II, there are around 22,500 raw alerts. We went through the same process steps as described in Section 7.1.1 to analyze and correlate alerts.

After alert aggregation and clustering, we got 1,800 hyper alerts. We also use the same network enclave used in Section 7.1.1 as an example to show our results in the GCP Scenario II.

In this network enclave, there are a total of 387 hyper alerts. Applying the *Ljung-Box* test to the hyper alerts, we identify 273 hyper alerts as the background alerts. In calculating the priority of hyper alerts, there are 9 hyper alerts whose priority values are above the threshold $\beta = 0.6$, meaning that we have more interest in these alerts than others.

As described in Section 6.1, we apply three correlation engines sequentially to the alert data to identify the alert relationship. For example, we select two alerts, *Plan_Service_Status_Down* and *Plan_Host_Status_Down*, as target alerts, then apply the GCT algorithm to correlating other alerts with them.

Table 7.6. Alert Correlation by the GCT on the GCP Scenario II: Target Alert: *Plan Service Status Down*

$Alert_i$	Target Alert	GCT Index
Plan_Registry_Modified	Plan_Service_Status_Down	20.18
HTTP_Java	Plan_Service_Status_Down	17.35
HTTP_Shells	Plan_Service_Status_Down	16.28

Table 7.7. Alert Correlation by the GCT on the GCP Scenario II: Target Alert: *Plan Server Status Down*

$Alert_i$	Target Alert	GCT Index
HTTP_Java	Plan_Server_Status_Down	7.73
Plan_Registry_Modified	Plan_Server_Status_Down	7.63
Plan_Service_Status_Down	Plan_Server_Status_Down	6.78
HTTP_RobotsTxt	Plan_Server_Status_Down	1.67

Table 7.6 and Table 7.7 show the corresponding GCT correlation results. In the tables, we list alerts whose *GCI* values have passed the *F*-test. The alerts *Plan_Host_Status* and *Plan_Service_Status* are issued by a network management system deployed on the network.

Figure 7.11 shows the correlation graph of *Plan Server*. The solid lines indicate the correct alert relationship while dotted lines represent false positive correlation. Figure 7.11 shows that *Plan_Registry_Modified* is causally related

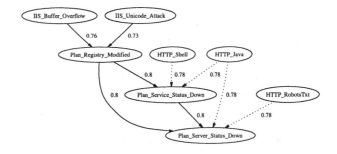

Figure 7.11. The GCP Scenario II: Correlation graph of the plan server

to alerts *Plan_Service_Status_Down* and *Plan_Server
_Status_Down*. The GCP document verifies such relationship. The attacker launched *IIS_Unicode_Attack* and *IIS_Buffer_Ove-rflow* attack against the Plan Server in order to traversal the root directory and access the plan server to install the malicious executable code. The Plan Server's registry file is modified (alert *Plan_Registry_Modified*) and the service is down (alert *Plan_Service_Status*) during the daemon installation. Alert *Plan_Host_Status_Down* indicates the "down" state of the plan server resulted from the reboot initiated by the malicious daemon. Plan server's states are affected by the activities of the malicious daemon installed on it. The ground truth described in the GCP document also supports the causal relationships discovered by our approach. In this experiment, the true positive correlation rate is 94.25% (vs. 93.15% using GCT-engine alone [46]) and false positive correlation rate is 8.92% (vs. 13.92% using GCT-engine alone [46]).

Table 7.8. Ranking of paths from node *IIS Buffer Overflow* to node *Plan Server Status Down*. $P = P(IIS\_Buffer\_Overflow)$

Order	Nodes Along the Path	Score
Path 1	IIS_Buffer_Overflow → Plan_Registry_Modified → Plan_Server_Status_Down	P* 0.61
Path 2	IIS_Buffer_Overflow → Plan_Registry_Modified Plan_Service_Status_Down	P*0.49

For nodes with multiple paths in the correlation graph, we can also perform path analysis quantitatively. For example, there are two paths connecting node *IIS_Buffer_Overflow* and node *Plan_Server_Status_Down* as shown in Figure 7.11. We can rank these two paths according to score of the overall likelihood, as shown in Table 7.8.

7.2.1   Discussion on GCP Scenario II.   Similar to our analysis in GCP Scenario I, our integrated correlation engine enables us to detect more cause-effect relationship between alerts. For example, in Figure 7.11, if using knowledge-based correlation engine, we can only detect the causal relationship between alerts *IIS_Buffer_Overflow* and *Plan_Registry_Modified*, as well as between alerts *IIS_Unicode_Attack* and *Plan_Registry_Modified*. With complementary temporal-based GCT alert correlation engine, we can detect other cause-effect relationship among alerts. For example, GCT-based correlation engine detected causality between a security alert (e.g.,*Plan_Registry_Modified*) and an alert output by the network management system (e.g., *Plan_Server_Status_Down*). In practice, it is difficult to detect such causality between security activity and network management fault using a knowledge-based correlation approach, unless such knowledge has been priory incorporated to the knowledge base.

Compared with GCP Scenario I, GCP Scenario II is more challenging due to the nature of the attack. Our correlation result in the GCP Scenario II is not comprehensive enough to cover the complete attack scenarios. By comparing the alert streams with the GCP document, we notice that many malicious activities in the GCP Scenario II are not detected by the IDSs and other security sensors. Therefore, some intermediate attack steps are missed, which is another challenge in GCP Scenario II.

Our approach depends on alert data for correlation and scenario analysis. When there is a lack of alerts corresponding to the intermediate attack steps, we cannot construct the complete attack scenario. In practice, IDSs or other security sensors can miss some attack activities. One solution is to apply attack plan recognition techniques that can partially link isolated attack correlation graphs resulted from missing alerts.

7.3   Discussion on Statistical and Temporal Correlation Engines

In our alert correlation system, we have designed three correlation engines. The Bayesian-based correlation aims to discover alerts that have direct causal relationship. Specifically, this correlation engine uses predicates to represent attack prerequisite and consequence, applies probabilistic reasoning to evaluating the property of *preparation-for relationship* between alerts. It applies time constraints to testing if the alert pair candidate conforms to the property of *sequential relationship* (i.e., causal alert appears before effect alert), and uses the pre-defined probability table of attack step transitions to evaluate the property of *statistical one-way dependence* (i.e., the probability that an effect alert occurs when a causal alert occurs) between alerts under correlation. Alert pairs that have matched these three properties are identified as having direct causal relationship.

In order to discover alerts that have no *known* direct causal relationship, we have also developed two statistical and temporal-based correlation models to discover *novel* and *new* attack transition patterns. The development of these two correlation techniques is based on the hypothesis that attack steps can still exhibit statistical dependency patterns (i.e., the third property of cause-effect alerts) or temporal patterns even though they do not have an obvious or *known* preparation-for relationship. Therefore, these two correlation engines aim to discover correlated alerts based on statistical dependency analysis and temporal pattern analysis with sequential time constraints. More formally, these two engines actually perform *correlation analysis* instead of a direct causality analysis because the preparation-for relationship between alerts are either indirect or unknown.

In theory, causality is a subset of correlation [24], which means that a causally related alert pair is also correlated, however, the reverse statement is not necessarily true. Therefore, the correlation output is actually a super set of correlated alerts that can include the causally related alert pairs as well as some correlated but non-causally related alerts. Our goal is to apply these two correlation engines to identifying the correlated alerts that have strong statistical dependencies and temporal patterns, and also conform to the sequential time constraint property. We present these correlated alert candidates to the security analysts for further analysis.

As an extra experiment, we applied GCP data sets to causal discovery-based correlation engine and GCT-based correlation engine only in order to test if the output of these two correlation engines can include the causally related alert pairs identified by Bayesian-based correlation engine. Our experiment results have shown that the correlated alerts identified by causal discovery-based correlation engine and GCT-based correlation engine have included those causally related alerts discovered by Bayesian-based correlation engine. In practice, we still use Bayesian-based correlation engine to identify causally related alerts in order to decrease the false positive correlation rate.

However, it does not necessarily mean that those two correlation engines (i.e., casual-discovery and GCT-based engines) can discover all the correlated alerts that have strong statistical and temporal patterns because of their limitations.

As described in Section 5.2, causal discovery-based correlation engine assumes that causality between variables can be represented by a causal Bayesian network that has a DAG structure. The statistical dependency between variables can be measured, for example, by mutual information. As described in Algorithm 1, causality direction among variables are identified by the assumption of causal Markov condition (i.e., a node X is independent with other nodes (except its direct effect nodes) given X's direct cause node) and the properties of *V-structure* as described in Section 5.2.2.

Due to the assumptions and properties used by causal discovery theory, in the process of alert correlation, the causal discovery-based correlation engine can result in cases that the causality direction cannot be identified among dependent alerts.

For example, for three variables A, B and C, after applying mutual information measures, we have got a dependency structure as $A - B - C$, which means A and B, B and C are mutually dependent respectively, A and C are mutually independent. If we apply conditional mutual information measure to A, B and C and get the result that A and C are conditionally independent given the variable B, then, without any other information, the causal discovery-based correlation engine actually cannot identify the causality among these three variables. In fact, with the above statistical dependency information, we can have the following three different causality structures, i.e., $A \rightarrow B \rightarrow C$, $A \leftarrow B \leftarrow C$ and $A \leftarrow B \rightarrow C$. These three causality structures have the same statistical dependency properties if no other information has been provided or extra causality has been identified (e.g., A or B or C has some dependency with another variable D, etc.). For the simplest dependency structure, i.e., $A - B$, without any extra information, causal discovery-based algorithm cannot identify the causality direction between A and B either.

By contrast, GCT-based correlation engine performs pairwise statistical dependency analysis and identify corresponding pairwise dependency direction. However, GCT-based correlation algorithm also incorporates the temporal information in the process of correlation. In particular, the GCT-correlation engine has the limitation of identifying correlated alert pairs whose time intervals have a *loose temporal pattern* (i.e., the variance of their time intervals has a large value) even though they may have a strong statistical dependency pattern.

In summary, considering the strength and limitations of causal discovery based and GCT-based correlation engines, from the perspective of statistical dependency and temporal pattern analysis, we can have a good correlation performance in identifying alerts that have a strong statistical dependency and strong temporal pattern because these two correlation engines can complement and enhance each other in this correlation space. If alerts that have a strong statistical dependency pattern but a loose temporal pattern, the correlation performance may be weak because GCT-based correlation engine has limitations in the loose temporal pattern space and causal discovery-based correlation engine also has its own limitations in the causality identification.

8.   Related Work

Recently, there have been several proposed techniques of alert correlation and attack scenario analysis.

Valdes and Skinner [52] proposed probabilistic-based approach to correlate security alerts by measuring and evaluating the similarities of alert attributes. In particular, the correlation process includes two phases. The first phase aggregates low-level events using the concept of attack threads. The second phase uses a similarity metric to fuse alerts into meta-alerts to provide a higher-level view of the security state of the system. Alert aggregation and scenario construction are conducted by enhancing or relaxing the similarity requirements in some attribute fields.

Porras et al. designed a "mission-impact-based" correlation system with a focus on the attack impacts on the protected domains [45]. The work is an extension to the prior system proposed in [52]. The system uses clustering algorithms to aggregate and correlate alerts. Security incidents are ranked based on the security interests and the relevance of attack to the protected networks and systems.

Some correlation research work are based on pre-defined attack scenarios or association between mission goals and security events. Goldman et al. [19] built a correlation system based on Bayesian reasoning. The system predefines the causal relationship between mission goals and corresponding security events as a knowledge base. The inference engine relies on the causal relationship library to investigate security alerts and perform alert correlation.

Debar and Wespi [14] applied backward and forward reasoning techniques to correlate alerts. Two alert relationships were defined, i.e., *duplicate* and *consequence*. In a correlation process, backward-reasoning looks for *duplicates* of an alert, and forward-reasoning determines if there are any *consequences* of an alert. They used clustering algorithms to detect attack scenarios and situations. This approach pre-defines consequences of attacks in a configuration file.

Krügel et al. [34] proposed a distributed pattern matching scheme based on an attack specification language that describes various attack scenario patterns. Alert analysis and correlation are based on the pattern matching scheme.

Morin and Debar [38] applied *chronicles formalism* to aggregating and correlating alerts. Chronicles provide a high level language to describe the attack scenarios based on time information. Chronicles formalism approach has been used in many areas to monitor dynamic systems. The approach performs attack scenario pattern recognition based on *known* malicious event sequences. Therefore, this approach is analogous to *misuse intrusion detection*.

Ning et al. [39], Cuppens and Miège [12] and Cheung et al. [7] built alert correlation systems based on matching the pre- and post-conditions of individual alerts. The idea of this approach is that prior attack steps prepare for later ones. Therefore, the consequences of earlier attacks correspond to the prerequisites of later attacks. The correlation engine searches alert pairs that have a consequence and prerequisite matching. In addition to the alert pre- and post-condition matching, the approach in [12] also has a number of phases including alert clustering, alert merging, and intention recognition. In the first two phases, alerts are clustered and merged using a similarity function. The intention recognition phase is referenced in their model, but has not been implemented. Having the correlation result, the approach in [39] further builds correlation graphs based on correlated alert pairs [39]. Recently, Ning et al. [41]

have extended the pre- and post-condition-based correlation technique to correlate some isolated attack scenarios by hypothesizing missed attack steps.

In the field of network management, alert or event correlation has been an active research topic and a subject of numerous scientific publications for over 10 years. The objective of alert correlation in a network management system (NMS) is to localize the faults occurred in communication systems. The problem of alert correlation in NMS is also referred as *root cause analysis*. During the past 10 more years, many solutions have been proposed, e.g., case-based systems [36], model-based approaches [42, 28] and code book-based technique [33]. The techniques are derived from different areas of computer science including artificial intelligence (AI), graph theory, neural networks, information theory, and automata theory.

Most of the proposed approaches have limited capabilities because they rely on various forms of predefined knowledge of attacks or attack transition patterns using attack modeling language or pre- and post-conditions of individual attacks. Therefore, those approaches cannot recognize a correlation when an attack is new or the relationship between attacks is new. In other words, these approaches in principle are similar to misuse detection techniques, which use the "signatures" of known attacks to perform pattern matching and cannot detect new attacks. It is obvious that the number of possible correlations is very large, potentially a combinatorial of the number of known and new attacks. It is infeasible to know *a priori* and encode all possible matching conditions between attacks. In practice, the more dangerous and intelligent adversaries will always invent new attacks and novel attack sequences. Therefore, we must develop significantly better alert correlation algorithms that can discover sophisticated and new attack sequences.

In the network management system (NMS), most event correlation techniques also depend on various knowledge of underlying networks and the relationship among faults and corresponding alerts. In addition, in an NMS, event correlation focuses more on alerts resulted from network faults that often have fixed patterns. Therefore, modeling-based or rule-based techniques are mostly applied in various correlation systems. Whereas in security, alerts are more diverse and unpredictable because the attackers are intelligent and can use flexible strategies. Therefore, it is difficult to apply correlation techniques developed in network management system to the analysis of security alerts.

Our approach aims to address the challenge of how to detect *novel* attack strategies that can consist of a series of unknown patterns of attack transitions. In alert correlation techniques, our approach differs from other work in the following aspects. Our approach integrates three complementary correlation engines to discover attack scenario patterns. It includes both knowledge-based correlation mechanisms and statistical and temporal-based correlation methods.

We apply a Bayesian-based correlation engine to the attack steps that are directly related, e.g., a prior attack enables the later one. Our Bayesian-based correlation engine differs from previous work in that we incorporate knowledge of attack step transitions as a constraint when conducting probabilistic inference. The correlation engine performs the inference about the correlation based on broad indicators of attack impacts without using the strict hard-coded pre-/post-condition matching.

In addition to domain knowledge-based correlation engine, we have developed two statistical and temporal-based correlation engines. The first one applies causal discovery theory to alert analysis and correlation. This approach identifies alert relationship based on statistical analysis of attack dependence. Having observed that many attack steps in a complicated attack strategy often have a strong temporal relationship, we have developed a correlation engine using temporal analysis. In particular, we applied Granger-Causality Test technique to discovering attack steps that have strong temporal and statistical patterns.

These two statistical and temporal-based correlation techniques differ from other related work in that they do not rely on prior knowledge of attack strategies or pre- and post-conditions of individual attacks. Therefore, these two statistical and temporal-based approaches can be used to discover *new* attack strategies that can have unknown attack transition patterns. To the best of our knowledge, our approach is the first approach that detects new attack strategies without relying on pre-defined knowledge base.

Our integrated approach also provides a quantitative analysis of the likelihood of various attack paths. With the aggregated correlation results, security analysts can perform further analysis and make inferences about high-level attack plans.

9. Conclusion and Future Work

In this paper, we have described an integrated alert correlation system designed to analyze INFOSEC alerts and detect novel attack strategies,

To meet the needs of detecting novel attack strategies, we have developed an integrated correlation system based on three complementary correlation techniques. Our correlation techniques are developed based on three hypotheses of attack step transitions. (1) The first hypothesis is that some attack steps have directly related connection, i.e., a prepare-for relationship. For this type of attack steps, we have developed a Bayesian-based correlation engine. It identifies alert causal relationship with a broad range of indicators of attack impacts. This correlation engine can also relax the strict hard-coded pre- and post-condition matching and handle the partial input evidence. (2) The second hypothesis is that some attack steps have statistical dependence patterns. We have developed and presented a statistical-based correlation engine based on causal discovery theory. (3) The third hypothesis is that attack steps have temporal patterns in their time intervals. For this type of attack relationship, we have built a correlation engine based on the Granger Causality Test. The major benefit provided by statistical and temporal correlation engines is that they can discover new attack transition patterns without relying on the domain knowledge.

We also described how to perform attack scenarios analysis by constructing correlation graphs based on correlation results. A quantitative analysis of attack strategy is conducted using the outputs of our integrated correlation engines. Attack strategies are analyzed using correlation graphs.

Finally, we have validated our correlation approach using DARPA Grand Challenge Problem (GCP) data set. The results have shown that our approach can effectively discover novel attack strategies with high accuracy.

In our future work, we will continue to study alert correlation with a focus on attack plan recognition and prediction.

References

[1] E. Bauer, D. Koller, and Y. Singer. Update rules for parameter estimation in Bayesian networks. In *Proceedings of the Thirteenth Conference on Uncertainty in Artificial Intelligence (UAI)*, pages 3–13, Providence, RI, August 1997.

[2] J. B. D. Cabrera, L. Lewis, X. Qin, W. Lee, and R.K. Mehra. Proactive intrusion detection and distributed denial of service attacks - a case study in security management. *Journal of Network and Systems Management*, vol. 10(no. 2), June 2002.

[3] J. B. D. Cabrera, L. Lewis, X. Qin, W. Lee, R. K. Prasanth, B. Ravichandran, and R. K. Mehra. Proactive detection of distributed denial of service attacks using mib traffic variables - a feasibility study. In *Proceedings of IFIP/IEEE International Symposium on Integrated Network Management (IM 2001)*, May 2001.

[4] J. B. D. Cabrera and R. K. Mehra. Extracting precursor rules from time series - a classical statistical viewpoint. In *Proceedings of the Second SIAM International Conference on Data Mining*, pages 213–228, Arlington, VA, USA, April 2002.

[5] P. E. Caines and C. W. Chan. Feedback between stationary stastic process. *IEEE Transactions on Automatic Control*, 20:495–508, 1975.

[6] J. Cheng, R. Greiner, J. Kelly, D. Bell, and W. Liu. Learning bayesian networks from data: An information-theory based approach. *Artificial Intelligence*, vol.137:43–90, 2002.

[7] S. Cheung, U. Lindqvist, and M. W. Fong. Modeling multistep cyber attacks for scenario recognition. In *Proceedings of the Third DARPA Information Survivability Conference and Exposition (DISCEX III)*, Washington, D.C., April 2003.

[8] R. V. Chitturi. Distribution of residual autocorrelations in multiple autoregressive schemes. *Journal of American Statistician Association*, 69:928–934, 1974.

[9] I. Cohen, A. Bronstein, and F. G. Cozman. Online learning of bayesian network parameters. *Hewlett Packard Laboratories Technical Report, HPL-2001-55(R.1)*, June 2001.

[10] G. F. Cooper and E. Herskovits. A bayesian method for constructing bayesian belief networks from databases. In *Proceedings of the Seventh Conference on Uncertainty in Artificial Intelligence*, 1991.

[11] T. Cover and J. Thomas. *Elements of Information Theory*. John Wiley, 1991.

[12] F. Cuppens and A. Miège. Alert correlation in a cooperative intrusion detection framework. In *Proceedings of the 2002 IEEE Symposium on Security and Privacy*, pages 202–215, Oakland, CA, May 2002.

[13] DAPRA Cyber Panel Program. DARPA cyber panel program grand challenge problem (GCP). http://www.grandchallengeproblem.net/, 2003.

[14] H. Debar and A. Wespi. The intrusion-detection console correlation mechanism. In *4th International Symposium on Recent Advances in Intrusion Detection (RAID)*, October 2001.

[15] D. A. Dickey and W. A. Fuller. Distribution of the estimators for autoregressive time series with a unit root. *Journal of American Statistician Association*, 74:427–431, 1979.

[16] N. Friedman, I. Nachman, and D. Peer. Learning bayesian network structure from massive datasets: The sparse candidate algorithm. In *Proceedings of the 15th Conference on Uncertainty in Artificial Intelligence*, 1999.

[17] C. W. Geib and R. P. Goldman. Plan recognition in intrusion detection system. In *DARPA Information Survivability Conference and Exposition (DISCEX II)*, June 2001.

[18] M. R. Gevers and B. D. O. Anderson. Representations of jointly stationary stochastic feedback processes. *International Journal of Control*, 33:777–809, 1981.

[19] R. P. Goldman, W. Heimerdinger, and S. A. Harp. Information modeling for intrusion report aggregation. In *DARPA Information Survivability Conference and Exposition (DISCEX II)*, June 2001.

[20] C. W. J. Granger. Investigating causal relations by econometric methods and cross-spectral methods. *Econometrica*, 34:424–428, 1969.

[21] IETF Intrusion Detection Working Group. Intrusion detection message exchange format. http://www.ietf.org/internet-drafts/draft-ietf-idwg-idmef-xml-09.txt, 2002.

[22] J. Haines, D. K. Ryder, L. Tinnel, and S. Taylor. Validation of sensor alert correlators. *IEEE Security & Privacy Magazine*, January/February, 2003.

[23] J. Hamilton. *Time Series Analysis*. Princeton University Press, 1994.

[24] A. J. Hayter. *Probability and Statistics for Engineers and Scientists*. Duxbury Press, 2002.

[25] D. Heckerman, C. Meek, and G. F. Cooper. A bayesian approach to causal discovery. In *Book of Computation, Causation, and Discovery, C. Glymour and G. Cooper, editors.* MIT Press, 1999.

[26] W. Hesse, E. Moller, M. Arnold, H. Witte, and B. Schack. Investigation of time-variant causal interactions between two eeg signals by means of the adaptive granger causality. *Brain Topography*, 15:265–266, 2003.

[27] J. R. M. Hosking. Lagrange multiplier tests of multivariate time series models. *Journal of The Royal Statistical Society Series B*, 43:219–230, 1981.

[28] G. Jakobson and M. Weissman. Real-time telecommunication network management: Extending event correlation with temporal constraints. In *Proceedings of the Fourth IFIP/IEEE International Symposium on Integrated Network Management (IM 1995)*, May 1995.

[29] S. Johansen. Statistical analysis of co-integration vectors. *Journal of Economic Dynamics and Control*, 1:321–346, 1988.

[30] K. Julisch and M. Dacier. Mining intrusion detection alarms for actionable knowledge. In *The 8th ACM International Conference on Knowledge Discovery and Data Mining*, July 2002.

[31] M. Kaminski, M. Ding, W. A. Truccolo, and S. L. Bressler. Evaluating causal relations in neural systems: Granger causality, direct transfer function (dtf) and statistical assessment of significance. *Biological Cybernetics*, 85:145–157, 2001.

[32] R. K. Kaufamnn and D. I. Stern. Evidence for human influence on climate from hemispheric temperature relations. *Nature*, 388:39–44, July 1997.

[33] S. Kliger, S. Yemini, Y. Yemini, D. Oshie, and S. Stolfo. A coding approach to event correlations. In *Proceedings of the 6th IFIP/IEEE International Symposium on Integrated Network Management*, May 1995.

[34] C. Krugel, T. Toth, and C. Kerer. Decentralized event correlation for intrusion detection. In *Proceedings of the 4th International Conference on Information Security and Cryptology*, 2001.

[35] H. Lee, K. S. Lin, and J. Wu. Pitfalls in using granger causality tests to find an engine of growth. *Applied Economics Letters*, 9:411–414, May 2002.

[36] L. Lewis. A case-based reasoning approach to the management of faults in communication networks. In *Proceedings of the IEEE INFOCOM*, 1993.

[37] G. M. Ljung and G. E. P. Box. On a measure of lack of fit in time series models. In *Biometrika 65*, pages 297–303, 1978.

[38] B. Morin and H. Debar. Correlation of intrusion symptoms: an application of chronicles. In *Proceedings of the 6th International Symposium on Recent Advances in Intrusion Detection (RAID 2003)*, Pittsburgh, PA, September 2003.

[39] P. Ning, Y. Cui, and D. S. Reeves. Constructing attack scenarios through correlation of intrusion alerts. In *9th ACM Conference on Computer and Communications Security*, November 2002.

[40] P. Ning and D. Xu. Learnign attack strategies from intrusion alerts. In *Proceedings of 10th ACM Conference on Computer and Communications Security (CCS'03)*, October 2003.

[41] P. Ning, D. Xu, C.G. Healey, and R. A. Amant. Building attack scenarios through integration of complementary alert correlation methods. In *Proceedings of the 11th Annual Network and Distributed System Security Symposium (NDSS'04)*, San Diego, CA, February 2004.

[42] Y. A. Nygate. Event correlation using rule and object based techniques. In *Proceedings of the 6th IFIP/IEEE International Symposium on Integrated Network Management*, May 1995.

[43] J. Pearl. *Probabilistic Reasoning in Intelligent Systems: Networks of Plausible Inference*. Morgan Kaufmann Publishers, Inc, 1988.

[44] J. Pearl. *Causality: Models, Reasoning, and Inference*. Cambridge University Press, 2000.

[45] P. A. Porras, M. W. Fong, and A. Valdes. A Mission-Impact-Based approach to INFOSEC alarm correlation. In *Proceedings of the 5th International Symposium on Recent Advances in Intrusion Detection (RAID)*, October 2002.

[46] X. Qin and W. Lee. Statistical causality analysis of INFOSEC alert data. In *Proceedings of the 6th International Symposium on Recent Advances in Intrusion Detection (RAID 2003)*, Pittsburgh, PA, September 2003.

[47] X. Qin and W. Lee. Attack plan recognition and prediction using causal networks. In *Proceedings of the 20th Annual Computer Security Applications Conference (ACSAC 2004)*, Tucson, AZ, December 2004.

[48] X. Qin and W. Lee. Discovering novel attack strategies from INFOSEC alerts. In *Proceedings of the 9th European Symposium on Research in Computer Security*, Sophia Antipolis, France, September 2004.

[49] S. M. Ross. *Introduction to Probability Models*. Harcourt Academic Press, 7th edition, 2000.

[50] P. Spirtes, C. Glymour, and R. Scheines. *Causation, Prediction, and Search*. Springer-Verlag NY, Inc., 1993.

[51] W. Stallings. *SNMP, SNMPv2, SNMPv3, and RMON 1 and 2*. Addison-Wesley, 1999.

[52] A. Valdes and K. Skinner. Probabilistic alert correlation. In *Proceedings of the 4th International Symposium on Recent Advances in Intrusion Detection (RAID)*, October 2001.

INDEX

Printed in the United States